ELECTRONIC PROPERTIES OF SEMICONDUCTING SOLID SOLUTIONS

ELEKTRONNYE SVOISTVA POLUPROVODNIKOVYKH TVERDYKH RASTVOROV

ЗЛЕКТРОННЫЕ СВОЙСТВА ПОЛУПРОВОДНИКОВЫХ ТВЕРДЫХ РАСТВОРОВ

Electronic Properties of Semiconducting Solid Solutions

by Aleksei B. Almazov
Department of Higher Mathematics
Moscow Mining Institute

Translated from Russian by Albin Tybulewicz
Editor, "Soviet Physics—Semiconductors"

Springer Science+Business Media, LLC 1968

Aleksei Borisovich 'Almazov, born in 1927, is Associate Professor in the Department of Higher Mathematics at Moscow Mining Institute and Research Scientist in the Chemistry Department at Moscow State University. A graduate of the Lenin Pedagogical Institute in Moscow, a former secondary school teacher, and for ten years a staff member at the Institute of Scientific Information and the Institute of General and Inorganic Chemistry of the Academy of Sciences of the USSR, Almazov received his Candidate's degree at Moscow State University in 1965. As a mathematical and theoretical physicist, Almazov has worked in polymer science, physico-chemical analysis, and solid state physics. As a scientific editor, from 1954 to 1965, he was responsible for the physical chemistry section of *Referativyi Zhurnal "Khimiya"* (the Soviet equivalent of *Chemical Abstracts*) and also edited and contributed to parts of the *Russian Encyclopedic Physics Dictionary*.

ISBN 978-1-4899-4838-0   ISBN 978-1-4899-4836-6 (eBook)
DOI 10.1007/978-1-4899-4836-6

The original Russian text, published by Nauka Press in Moscow in 1966 for the N. S. Kurnakov Institute of General and Inorganic Chemistry of the Academy of Sciences of the USSR, has been corrected by the author for this edition.

Алексей Борисович Алмазов

Электронные свойства полупроводниковых твердых растворов

Library of Congress Catalog Card Number 68-18820

PREFACE TO THE AMERICAN EDITION

It is with great pleasure that I learned that my book would appear in an English edition. The interest in semiconducting solid solutions — even the interest in my fragmentary presentation in this monograph — is understandable: in many applications, these materials are likely to replace soon the stoichiometric crystals because their properties can be changed gradually by a continuous variation of their composition, while retaining at the same time many features of the "classical" semiconductors (long lifetimes, high mobilities, etc.). The main difficulty in the application of solid solutions is the complex technology of preparation in a form suitable for further use. However, the problem is more economic than scientific. The main and supplementary lists of references are not exhaustive: the cited papers should be regarded as reference points in the jungle of published literature.

It will give me pleasure if, in addition to its scientific purpose, which is to stimulate interest in this promising branch of solid state physics, this monograph will help in strengthening scientific and personal contacts between Soviet scientists and their colleagues abroad.

A. B. Almazov

September 27, 1967

CONTENTS

CONTENTS

INTRODUCTION

The currently favored theory of crystals is based on the assumption of an ideal periodic lattice, consisting of a regular array of unit cells free of distortions and defects. This assumption makes it possible to formulate certain basic concepts: crystal momentum, which is a vector quantity representing any nondecaying excited state of a crystal; a zone, which is an assembly of states transformed in accordance with the translation group representations, etc. These concepts are then used to describe slight departures from perfect periodicity, such as weak vibrations of the lattice, impurities present in low concentrations, low densities of dislocations, etc. This approach cannot be used to deal with "large" departures from the ideal state.

The departure from the ideal state need not be very large for the aforementioned concepts to lose their meaning. At the same time, many derived concepts also lose their meaning. Thus, the band structure representation becomes meaningless because of the absence of the crystal momentum. It is also known that at sufficiently high impurity concentrations in a semiconductor, a "tail" appears in the density of states, and therefore the concept of a forbidden band becomes incorrect, so that even the fundamental division of ideal crystals into metals, on the one hand, and semiconductors and dielectrics, on the other, loses its meaning: the same sample can be regarded either as a metal or as a semiconductor. These observations indicate that the theory of solids which differ considerably from ideal crystals should be based on a completely different principle.

A theory considering deviations from the ideal state which are of arbitrary magnitude is useless because of the extreme generality of the initial assumptions. It would therefore seem desirable to consider systems which are different from ideal crystals in only certain respects. Among such systems are solid solutions, which retain a regular array of points at which atoms are located or may be located, but in which it is no longer assumed that a given site in a unit cell is occupied exclusively by an atom of a definite kind. Thus, solid solutions are similar to ideal crystals in one respect: the atoms are located at sites of a regular crystal lattice. The difference is that the atoms of the components of a solution are distributed at random in the sites, whereas, in an ideal crystal, the distribution is orderly.

A substitutional solution differs from an ideal crystal only because a site may contain an atom of one or the other component; an interstitial solution differs only because atoms are located in interstices, which form a regular sublattice. If we use the traditional representations of the elementary theory of probability, then the simplest model of a binary solution is an ideal lattice, the sites of which are occupied by randomly distributed black and white balls. The equivalent sites of such a lattice differ from one another in that there is a definite probability of finding a black or a white ball at a given site, so that when we move along a given crystallographic direction, we observe a certain statistical sequence of two alternative events (i.e., finding a black or a white ball). Such a system represents a process with discrete (vector) time and independent values of the relevant quantities; a step in this process is a transition from one site to its neighbor, and the finding of a black or a white ball is one of the two alternative events.

This model suggests that the correct system of concepts, suitable for the description of the properties of solid solutions, is provided by the theory of probability processes with lattice distribution functions. Using this formulation of the problem, we shall consider only those solid solutions which are thermodynamically stable nonstoichiometric uniform solids. The components of a solid solution must usually have similar chemical properties and, in particular, should be in the same group of the periodic system. Thus, semiconductors doped with chemically different elements (which have been investigated thoroughly both theoretically and experimentally) are outside our scope because such doping usually results not in a solid solution but in a system which decomposes into two phases: the host matrix and the dopant. The fact that the decomposition may take a very long time is of no importance; all that is important is that the property—composition diagrams of such systems usually have no features characteristic of solid solutions.

When a semiconductor is doped with chemically different elements, local impurity states are produced. These states appear because donor or acceptor impurities produce Coulomb fields sufficient for the formation of fairly deep potential wells (traps). In a solid solution, such fields do not arise because of the chemical similarity of the components, which have the same number of valence electrons; therefore, impurity states are not observed, at least at normal temperatures. Consequently, departures from the ideal crystal potential, due to the partial replacement of an element with a chemically similar one, are slight; this characteristic will be exploited systematically in several variants of the perturbation theory. This does not mean that we shall ignore solid solutions which exhibit ionic binding and consequently have Coulomb fields. For our purpose it will be important that, in the presence of such fields, the potential — in which carrier generation and motion take place — should be capable of being approximated by the potential of an ideal crystal. A considerable part of the present monograph is devoted to the elucidation of the meaning of this approximation.

Semiconducting solid solutions are interesting because of their possible applications. Thus, Ge—Si solutions retain many properties of ideal crystals, and some of these properties vary continuously with the composition. This makes it possible to prepare semiconductors with continuously varying properties and thus obtain the required values of certain parameters. Solutions of $A_2^V B_3^{VI}$ compounds are interesting because of their thermoelectric properties. It follows from the most general considerations [57, 58] that such solid solutions should have good thermoelectric properties: under certain conditions, the carrier mobility in a solution may be close to the mobility in an ideal crystal, while phonons will be damped because of departures from the periodicity. Thus, a considerable fraction of heat will be transported by carriers. Solid solutions may be useful in laser technology because of the possibility of varying continuously the properties (particularly the transition energies) by varying the composition.

The author's own work has been concerned with the behavior of fermions in random fields [6, 9, 16, 17]. The first sections present the probability theory representations, which must be used to deal with the electronic properties of solid solutions, and an explanation of the relationship between the concepts introduced and the concepts in the theory of ideal crystals. In these sections we shall use, without any further reference, the methods and results of the theory of lattice distributions [27, 73] and of the theory of probability processes [12]. The generally accepted representation that carriers themselves form a spectrum will also be regarded as an important concept. It will be shown that the appearance of free carriers at higher temperatures screens and smooths out the random field, and this effect can be observed experimentally. The situation is in many respects similar to the screening observed in classical phenomena. Thus, solvated ions in electrolytes take part in all the phenomena and "bare" charges do not appear at all.

The review part of the monograph presents the published theoretical and experimental data on the electronic properties of solid solutions and lays down the principles to be followed in the interpretation of the experimental data on the basis of the proposed theory. The experimental data have been gathered mainly from several reviews [155, 168, 229].

For the sake of simplicity, the treatment in Part I is restricted to a binary solid solution having a unipolar n-type or p-type conduction.

For certain reasons (cf., for example, [15]), even a theory of the electronic properties of ideal crystals cannot be developed axiomatically, and at some stage it is necessary to use the experimental data and descriptive treatment. This approach is adopted here also, because random fields complicate the problem.

The following Fourier transformation rules are used:

$$f(\mathbf{x}) = \int d\mathbf{k} e^{i\mathbf{k}\mathbf{x}} f(\mathbf{k}), \quad f(\mathbf{k}) = \frac{1}{(2\pi)^3} \int d\mathbf{x} e^{-i\mathbf{k}\mathbf{x}} f(\mathbf{x}).$$

PART I

THEORY OF THE ELECTRONIC PROPERTIES OF A SEMICONDUCTING SOLID SOLUTION

1. PROBABILITY CHARACTERISTICS OF A RANDOM POTENTIAL IN A SOLID SOLUTION

First, we shall assume that the potential in which a carrier is traveling in a solid solution is the sum of the potentials of force centers located at the sites of a regular crystal lattice. Secondly, we shall assume that these force centers occupy the lattice sites in a statistically independent manner, so that the probability of a given lattice site being occupied by a particular atom is independent of the nature of the atoms at other sites. The latter assumption is, of course, implicit in the definition of a disordered solid solution; if there were any correlation between the positions of atoms, a solid solution would decompose into a two-phase system or at least local deviations from stoichiometry would appear in the sample (we shall ignore these deviations). The first assumption is closely related to the second. We shall divide a crystal in the usual manner into cells, so that each cell contains an atom of one or the other kind, and the whole volume is filled with cells. We shall assume that the potential due to nuclei and electrons located in a single cell is the potential of a force center. If the potential of any one center were to depend on the potentials of neighboring centers, then there would be a correlation between the positions of the atoms, resulting from a redistribution of the electron density. The treatment given here can be regarded as the first approximation to a more rigorous theory, in which the first assumption is not satisfied.

Let us assume that $\alpha(\mathbf{X})$ are random numbers which can assume the values +1 and −1 with probabilities n_1 and n_2 independent of $\mathbf{X}$, and that $n_1 + n_2 = 1$; $\mathbf{X}$ represents vectors in a sublattice whose sites are occupied by the components of the solution. The symmetry of this sublattice need not be identical with the symmetry of crystals formed from the pure components. The probabilities n_1 and n_2 are the ratios of the numbers of atoms of one or the other kind to the total number of all atoms, i.e., they are dimensionless concentrations (in the case of interstitial solid solutions, they represent the concentrations of empty and occupied interstices).

The potential established at a point $\mathbf{x}$ by an atom (of one kind or the other) located at a site $\mathbf{X}$ is a random quantity related to $\alpha(\mathbf{X})$ by

$$V(\mathbf{x},\mathbf{X}) = v(\mathbf{x} - \mathbf{X}) + \frac{1}{2}\alpha(\mathbf{X})w(\mathbf{x} - \mathbf{X}),$$
$$v(\mathbf{x}) = \frac{1}{2}[v_1(\mathbf{x}) + v_2(\mathbf{x})],$$
$$w(\mathbf{x}) = v_1(\mathbf{x}) - v_2(\mathbf{x}). \tag{1}$$

Here, $v_1(\mathbf{x})$ and $v_2(\mathbf{x})$ are potentials due to atoms of one kind or the other (in the case of interstitial solutions, one of these potentials is taken to be zero). These potentials will be assumed to be arbitrary within physically reasonable limits. The purpose of this part of our treatment is simply to establish the general pattern of the behavior of carriers in semiconducting solid solutions. It follows from Eq. (1) that $V(\mathbf{x}, \mathbf{X})$ is a random quantity, assuming the value $v_1(\mathbf{x} - \mathbf{X})$ if $\alpha(\mathbf{X}) = +1$ and $v_2(\mathbf{x} - \mathbf{X})$ if $\alpha(\mathbf{X}) = -1$.

A characteristic function of the quantity $V(\mathbf{x}, \mathbf{X})$ is defined as

$$f_{\mathbf{x}}(s) = \mathrm{P}\{\alpha(\mathbf{X}) = 1\}\, e^{isv_1(\mathbf{x}-\mathbf{X})} + \mathrm{P}\{\alpha(\mathbf{X}) = -1\}\, e^{isv_2(\mathbf{x}-\mathbf{X})},$$

where $\mathrm{P}\{\ldots\}$, which is the probability of the event described in the curly brackets (braces), has the form

$$f_{\mathbf{x}}(s) = n_1 e^{isv_1(\mathbf{x}-\mathbf{X})} + n_2 e^{isv_2(\mathbf{x}-\mathbf{X})}. \tag{2}$$

The potential established at a point $\mathbf{x}$ by all atoms of the solid solution is, by definition, the sum of the potentials of individual atoms:

$$V(\mathbf{x}) = \Sigma V(\mathbf{x}, \mathbf{X}). \tag{3}$$

From the independence of the random quantities $\alpha(\mathbf{X})$, it follows that the characteristic function $f(s)$ of the sum $V(\mathbf{x})$ of independent terms is equal to the product of the characteristic functions of the type given by Eq. (2):

$$f(s) = \Pi\left(n_1 e^{isv_1(\mathbf{x}-\mathbf{X})} + n_2 e^{isv_2(\mathbf{x}-\mathbf{X})}\right). \tag{4}$$

In Eqs. (3) and (4), and in every subsequent case, we shall assume, unless otherwise stated, that the summation and multiplication is carried out over all values of $\mathbf{X}$.

The distribution function $F'(V)$ [for the sake of simplicity, we shall omit the argument of $V(\mathbf{x})$] is found using the inversion formula

$$F'(V) = \frac{1}{2\pi} \int_{-\infty}^{\infty} e^{-isV} f(s)\, ds. \tag{5}$$

A cumulant function

$$\psi(s) = \ln f(s) = \Sigma \ln \left(n_1 e^{isv_1(\mathbf{x}-\mathbf{X})} + n_2 e^{isv_2(\mathbf{x}-\mathbf{X})}\right)$$

can be employed in the usual manner to obtain semi-invariants and moments. If the cumulant function is differentiable p times, it can be represented in the form

$$\psi(s) = \sum_{k=1}^{p} \frac{(is)^k}{k!} \gamma_k + O(s),$$

where $O(s)$ is a nondifferentiable function. The quantities γ_k are known as semi-invariants; they can be rationally expressed in terms of moments of a random quantity (in this case, the potential at a point $\mathbf{x}$) and, conversely, the moments can be expressed in terms of semi-invariants, where γ_1 is simply a mathematical expectation and γ_2 is the variance. Thus, the mathematical expectation of the potential and its variance at a point $\mathbf{x}$ are equal:

$$\mu(\mathbf{x}) = \mathbf{M}V(\mathbf{x}) = n_1 \Sigma v_1(\mathbf{x} - \mathbf{X}) + n_2 \Sigma v_2(\mathbf{x} - \mathbf{X}) \equiv n_1 V_1(\mathbf{x}) + n_2 V_2(\mathbf{x}),$$

$$\sigma^2(\mathbf{x}) = n_1 n_2 \Sigma w^2(\mathbf{x} - \mathbf{X}), \tag{6}$$

$$V_1(\mathbf{x}) = \Sigma v_1(\mathbf{x} - \mathbf{X}), \quad V_2(\mathbf{x}) = \Sigma v_2(\mathbf{x} - \mathbf{X})$$

(**M** denotes the mathematical expectation).

We shall now consider the properties of the distribution function (5). If the potentials $v_1(\mathbf{x})$ and $v_2(\mathbf{x})$ are not equal to zero in some finite region, the distribution function is degenerate. In this case, the characteristic function is the sum of a finite number of exponential functions and the distribution function is the sum of a finite number of δ functions. Let us assume that, for example, $v_1(\mathbf{x})$ and $v_2(\mathbf{x})$ are equal to zero outside a cell in which the given atom is found. Then

$$F'(V) = n_1 \delta[V - v_1(\mathbf{x})] + n_2 \delta[V - v_2(\mathbf{x})]. \tag{7}$$

Thus, in the model considered, we have two alternatives: either the potentials $v_1(\mathbf{x})$ and $v_2(\mathbf{x})$ are different from zero in some finite region and then the distribution function represents a convolution of a finite number of degenerate distributions, or the potential at a given point includes contributions of all atoms and then the distribution is a convolution of an infinite number of terms. The probability properties of such sums can be considered separately for each case by proving specific limit theorems.

We shall show that at low concentrations of one of the components the characteristic function (4) reduces to a characteristic function which is used in the theory of moderately doped semiconductors [17]. For this purpose, we shall replace n_1 and n_2 with their asymptotic values N_1/N and N_2/N, where N_1 and N_2 are the numbers of atoms of the first and second components, and N is the total number of atoms, $N_1 + N_2 = N$. We shall assume the concentration of one of the components to be so small that $N_2 \ll N_1 \approx N$ and asymptotically $N_2/\omega = n_2' < \infty$, where ω is a normalizing volume. When these assumptions are made, we find that

$$\ln f(s) = n_2' \int d\mathbf{x} \, (e^{isv_2(\mathbf{x})} - 1),$$

which is indeed identical with the cumulant function found in [17].

The distribution function (5) is, in accordance with the spirit of the problem, translationally invariant: if $\mathbf{X}$ is altered by an amount equal to any lattice vector, the characteristic function $f(s)$ and the distribution function $F(V)$ are not affected.

The parametric dependence of V on $\mathbf{x}$ justifies the treatment of this quantity as a random function or, if it is convenient, as a random process with a vector time $\mathbf{x}$. For simplicity, we shall introduce a centered random function

$$V'(\mathbf{x}) = V(\mathbf{x}) - \mu(\mathbf{x}),$$

which obviously has the property $\mathbf{M}V'(\mathbf{x}) = 0$. We shall be interested in the correlation properties of the function $V'(\mathbf{x})$. To simplify the problem further, we shall alter somewhat the standard definition of the correlation function, so that it becomes

$$R\left(\mathbf{X},\ \mathbf{Y}\right)=\frac{1}{\Omega^2}\int_{\Omega\left(\mathbf{X}\right)}d\mathbf{x}\int_{\Omega\left(\mathbf{Y}\right)}d\mathbf{y}MV'\left(\mathbf{x}+\mathbf{X}\right)V'\left(\mathbf{y}+\mathbf{Y}\right),\tag{8}$$

where the integration is carried out over the volumes Ω of cells, whose positions are given by the lattice vectors $\mathbf{X}$ and $\mathbf{Y}$ (these volumes are all equal to Ω). Because of the translational invariance of the mathematical expectation, R($\mathbf{X}$, $\mathbf{Y}$) depends only on the difference $(\mathbf{X} - \mathbf{Y})$. Moreover, we may assume that if we are dealing with a crystal, whose "average" cells have inversion centers, then the correlation function R($\mathbf{X} - \mathbf{Y}$) depends only on the absolute value of each of the components of the vector $(\mathbf{X} - \mathbf{Y})$. Thus, R($\mathbf{X} - \mathbf{Y}$) is a correlation function of a stationary (for each component) process. This circumstance simplifies considerably our treatment without affecting greatly the physical meaning: integration in accordance with Eq. (8) means, in fact, that we are neglecting distances (wavelengths) of the order of the lattice constant; carriers of such wavelengths are practically never found in a crystal and the influence of potential wells or "hillocks" extending over such distances can be easily allowed for parametrically, as shown later. Since

$$M\alpha^2\left(\mathbf{X}\right)=1,$$

$$M\alpha\left(\mathbf{X}\right)=n_1-n_2$$

and $\alpha(\mathbf{X})$ are independent for different $\mathbf{X}$, the correlation function for the quantities $\alpha(\mathbf{X})$ has the form

$$M\alpha\left(\mathbf{X}\right)\alpha\left(\mathbf{Y}\right)=\left(n_1-n_2\right)^2+4n_1n_2\delta\left(\mathbf{X}-\mathbf{Y}\right).\tag{9}$$

We shall now calculate the correlation function (8). Since

$$V'\left(\mathbf{x}\right)=\frac{1}{2}\Sigma\left[\left(n_1-n_2\right)+\alpha\left(\mathbf{X}\right)\right]w\left(\mathbf{x}-\mathbf{X}\right),$$

it follows that

$$MV'\left(\mathbf{x}+\mathbf{X}\right)V'\left(\mathbf{y}+\mathbf{Y}\right)=n_1n_2\sum_{\mathbf{Z}}w\left(\mathbf{x}+\mathbf{Z}\right)w\left(\mathbf{y}+\mathbf{Z}+\mathbf{X}-\mathbf{Y}\right).$$

Using Eq. (9), we find that

$$R\left(\mathbf{X}-\mathbf{Y}\right)=\frac{n_1n_2}{\Omega^2}\sum_{\mathbf{Z}}\int_{\Omega\left(\mathbf{X}\right)}d\mathbf{x}\int_{\Omega\left(\mathbf{Y}\right)}d\mathbf{y}w\left(\mathbf{x}+\mathbf{Z}\right)w\left(\mathbf{y}+\mathbf{Z}+\mathbf{Y}-\mathbf{X}\right).$$

Hence,

$$R\left(\mathbf{X}\right)=\frac{n_1n_2}{\Omega^2}\int_{\Omega\left(0\right)}d\mathbf{x}\,d\mathbf{y}\sum_{\mathbf{Y}}w\left(\mathbf{x}+\mathbf{Y}\right)w\left(\mathbf{y}+\mathbf{X}+\mathbf{Y}\right).\tag{10}$$

As expected,

$$R\left(0\right)=\sigma_0^2=\frac{1}{\Omega^2}\int_{\Omega}\sigma\left(\mathbf{x}\right)\sigma\left(\mathbf{y}\right)d\mathbf{x}\,d\mathbf{y},\tag{11}$$

and it follows from the stationary nature of the problem that R(0) $\geq$ R($\mathbf{X}$).

We have already mentioned that the properties of V($\mathbf{x}$) depend strongly on whether the potentials $v_1(\mathbf{x})$ and $v_2(\mathbf{x})$ vanish at some finite distance or at infinity. We shall find how this dependence is reflected in the properties of the correlation function R($\mathbf{X}$). It follows directly from Eq. (2) that if w($\mathbf{x}$) differs from zero within a certain finite region, then the sum (10) consists only of a finite number of terms because the remainder vanishes. Otherwise, the potential at any point includes contributions of all atoms and R($\mathbf{X}$) is an infinite series.

In addition to R($\mathbf{X}$), we can consider its Fourier transform

$$\sum e^{iKX} R(\mathbf{X}) \equiv \sigma_0^2 R(\mathbf{K}), \tag{12}$$

where the vector $\mathbf{K}$ is defined in the usual manner in the first Brillouin zone and R($\mathbf{K}$) is periodic in the $\mathbf{K}$ space.

In fact, the addition to $\mathbf{K}$ of any reciprocal lattice vector does not alter R($\mathbf{K}$) because of the periodicity of the exponential function. Therefore, all the results of the Brillouin zone theory apply automatically to our treatment. In particular, we can speak of weak correlation, which is the analog of the "nearest neighbor approximation," when R($\mathbf{K}$) vanishes at distances larger than the radius of the first coordination sphere, so that the sum in Eq. (12) consists only of terms corresponding to the first coordination sphere; or we can consider strong correlation, when all R($\mathbf{X}$) tend to the same constant value (which is the analog of the "almost-empty lattice approximation"). We can also use an interpolation scheme, which covers both cases [7]. We can assume that

$$R(\mathbf{X}) = \sigma_0^2 e^{-X^2/a^2}, \tag{13}$$

where a is a parameter with the dimensions of length, and with an obvious physical meaning; at distances considerably larger than a, R($\mathbf{X}$) practically vanishes so that the strong and weak correlation apply, respectively, when $a \to \infty$ and $a \to 0$. Then, as shown by direct calculation in [7], the summation over the lattice sites gives the following expressions. If the sites or interstices occupied by the second component form a simple cubic lattice, then

$$R(\mathbf{K}) = \vartheta_3(K_1)\,\vartheta_3(K_2)\,\vartheta_3(K_3). \tag{14}$$

Similarly, for a body-centered cubic lattice, we have

$$R(\mathbf{K}) = \vartheta_3(K_1)\,\vartheta_3(K_2)\,\vartheta_3(K_3) + \vartheta_2(K_1)\,\vartheta_2(K_2)\,\vartheta_2(K_3), \tag{15}$$

and for a face-centered cubic lattice,

$$R(\mathbf{K}) = \vartheta_3(K_1)\,\vartheta_3(K_2)\,\vartheta_3(K_3) + \vartheta_3(K_1)\,\vartheta_2(K_2)\,\vartheta_2(K_3) + \vartheta_2(K_1)\,\vartheta_3(K_2)\,\vartheta_2(K_3) + \vartheta_2(K_1)\,\vartheta_2(K_2)\,\vartheta_3(K_3). \tag{16}$$

In formulas (14)-(16), we have used the standard notation for the ϑ functions:

$$\vartheta_3(K_i;\ v) \equiv \vartheta_3(K_i) = 1 + 2\sum_{n=1}^{\infty} e^{-\pi v n^2} \cos 2n K_i,$$

$$\vartheta_2(K_i;\ v) \equiv \vartheta_2(K_i) = 2\sum_{n=1}^{\infty} e^{-\pi v (n-\frac{1}{2})^2} \cos(2n-1) K_i.$$

the K_i denote the components of the reciprocal lattice vector, multiplied by the lattice constant d. The parameter ν, which occurs in the definition of the ϑ functions, is related to d by

$$\pi\nu = d^2/a^2.$$

Thus, the strong correlation case corresponds to $\nu \to 0$, and the weak correlation, to $\nu \to \infty$.

In the theory of the elliptic functions, a different characteristic of the ϑ functions is used: this characteristic is the "modular angle," which is in one-to-one correspondence with ν. We can show that the strong and weak correlation correspond, respectively, to modular angles close to 90° and 0°.

In addition to their use in an interpolation method as described here, the ϑ functions also have the following property. Let us assume that we wish to define a complex crystal momentum $\mathbf{K} = \mathbf{K'} + i\mathbf{K''}$ so that the function $R(\mathbf{K})$ is periodic with respect to the vector $\mathbf{K'}$, i.e., so that it does not change when $\mathbf{K'}$ is replaced with $\mathbf{K'} + \mathbf{K_l}$, where $\mathbf{K_l}$ is any reciprocal lattice vector, and, moreover,

$$|R(\mathbf{K})| = e^{-(\rho\mathbf{K''})}|R(\mathbf{K'})|,$$

where ρ is some vector. We shall show in Section 3 that in some problems $R(\mathbf{K})$ can be regarded as simply proportional to the scattering cross section. From this point of view, the requirement that $R(\mathbf{K})$ should decay exponentially — when a complex crystal momentum is introduced — is analogous to a similar requirement in the case of a plane wave in a continuous medium: the amplitude of such a wave may be written in the form

$$U(r, t) = U_0 e^{i\omega t + i(pr)} = U_0 e^{-(p''r)} e^{i\omega t + i(p'r)},$$

where $p = p' + ip''$, which makes it possible to express the absorption coefficient in terms of p''. It is shown in the general theory of ϑ functions [118a, 198a] that among the integral functions only the ϑ functions are simultaneously periodic along the real axis and exponentially damped if the argument is complex. Thus, the ϑ functions play the same role in the lattice theory as the exponential functions in the theory of continuous media.

The limiting case of the total absence of correlation (a process with independent increments) is obtained using the distribution functions (7) and, as can be deduced directly from Eqs. (10) and (12),

$$R(X) = \sigma_0^2 \delta(X), \quad R(K) = 1. \tag{17}$$

We can also calculate correlation functions of higher orders, describing the correlation of the potential at three or more points. The electronic properties of a solid solution are defined by a sequence of correlation functions — in the simplest case, by one correlation function, (10) is used.

The explicit dependence of the probability characteristics of the random potential in a solid solution on the lattice structure (discussed in the present section) is required in those problems where such a dependence is known to affect the investigated quantities (for example, in the discussion of the band structure and similar problems). However, even in the theory of ideal crystals these problems have not yet been solved finally. Thus, because of the explicit dependence of the effective wave equation on temperature it is not clear to what degree we can regard the band structure as "frozen," i.e., independent of temperature. When these general problems are solved in the theory of ideal crystals, the general relationships obtained can then be applied to the theory of electronic properties of solid solutions.

We shall now consider a simplified model, which is the analog of the effective mass approximation.

2. STATIONARY FIELDS AND THE SCREENING OF THESE FIELDS BY CARRIERS

In many problems of the theory of semiconductors, we can ignore the periodic field in a crystal, assuming electrons or holes to be free particles whose kinetic energy ε depends on the momentum $\hbar k$, as if the crystal were altogether absent:

$$\varepsilon(k) = \frac{\hbar^2 \mathbf{k}^2}{2m}. \tag{18}$$

In the presence of several minima in a band, we use a sum of expressions of the type (18); such generalization introduces nothing new. In fact, the approximation (18) implies the neglecting of wavelengths comparable with the lattice constant, by expanding the dispersion law as a series near a minimum. This assumption is justified, since we are not interested in high energies. If we neglect distances of the order of the lattice constant, it means that, for example, in the formulas of the preceding section the lattice vectors should be assumed to be continuously variable and sums should be replaced with integrals. Thus, as in the case of Eq. (18), where we assume some fictitious uniform and isotropic medium (allowing parametrically for the presence of a crystal using an effective mass) in our problem of random fields, we must go over to statistically uniform fields. The properties of such fields have been investigated in detail [12]. In particular, it is known that if $\varphi(\mathbf{x})$ is a statistically uniform field, this field can be represented in the form

$$\varphi(\mathbf{x}) = \int e^{i\mathbf{k}\mathbf{x}} \varphi(\mathbf{k}) \, d\mathbf{k}, \tag{19}$$

where $\varphi(\mathbf{k})$ is another field defined in the $\mathbf{k}$ space and having the property that $M\{\varphi(\mathbf{k}) d\mathbf{k} \varphi(\mathbf{k'}) \cdot d\mathbf{k'}\} = 0$ in all cases whenever the regions $d\mathbf{k}$ and $d\mathbf{k'}$ do not overlap, i.e.,

$$M\{\overline{\varphi}(\mathbf{k}) \varphi(\mathbf{k'})\} = \rho(\mathbf{k}) \delta(\mathbf{k} - \mathbf{k'}), \tag{20}$$

where the bar denotes the complex conjugate. The function $\rho(\mathbf{k})$ is known as the spectral density of the field $\varphi(\mathbf{x})$. In Eqs. (19) and (20), and elsewhere, the points in space are denoted by small letters, in order to stress that we are dealing with a continuous medium. The field, in addition to being stationary, must be statistically isotropic, which limits us to crystals of the cubic symmetry.

In general, the expressions $\varphi(\mathbf{k})d\mathbf{k}$ in formulas (19) and (20) cannot be regarded as differentials of some function, since not every random function is differentiable. Therefore, for example, Eq. (19) should be regarded either as a Stieltjes integral or $\varphi(\mathbf{k})$ should be regarded as a generalized function.

It is known [15] that, in the many-electron approach, the behavior of carriers in semiconductors is self-consistent: the spectrum of carriers is governed not only by external fields, but also by the carriers themselves. In other words, the potential in which a carrier moves depends on the presence or absence of other carriers (i.e., on temperature), and in that sense carriers "form their own spectrum." This effect is manifested primarily in the screening of external fields. We shall consider how such screening affects the probability characteristics of external fields. We shall assume that we have some centered "primary" stationary and isotropic field with a correlation function

$$\mathbf{M}\varphi(\mathbf{x})\,\varphi(\mathbf{x}') = \rho(\mathbf{x} - \mathbf{x}').\qquad(21)$$

Using the general theory of stationary processes, we can show that the right-hand part of Eq. (21) depends only on the difference $(\mathbf{x} - \mathbf{x}')$. The problem is to calculate the probability characteristics of a screened field $\widetilde{\varphi}(\mathbf{x})$, in particular, the correlation function of this field.

An improved version of the perturbation theory [15] gives the following expression for the screened field:

$$\widetilde{\varphi}(\mathbf{x}) = \int \frac{e^{i\mathbf{k}\mathbf{x}}\,\varphi(\mathbf{k})\,d\mathbf{k}}{1 - (2\pi)^4\,\mathbf{k}^{-2}\,P(\mathbf{k},\,0)}.\qquad(22)$$

Here, $P(\mathbf{k}, 0)$ is a polarization operator; explicit expressions for this operator in cases of interest to us will be given later.

Strictly speaking, the polarization operator in the problem considered is a functional of the random field. By way of approximation, we shall assume that $P(\mathbf{k}, 0)$ is not a random quantity but simply a number, while the parameters on which $P(\mathbf{k}, 0)$ depends (such as the Debye radius, etc.) can be regarded as averages.

Using the property described by Eq. (20), we find

$$\widetilde{\rho}(\mathbf{x}) = \int \frac{e^{i\mathbf{k}\mathbf{x}}\,\rho(\mathbf{k})\,d\mathbf{k}}{[1 - (2\pi)^4\,\mathbf{k}^{-2}\,P(\mathbf{k},\,0)]^2},\qquad(23)$$

and hence we see immediately that the screened spectral density

$$\widetilde{\rho}(\mathbf{k}) = \frac{\rho(\mathbf{k})}{[1 - (2\pi)^4\,\mathbf{k}^{-2}\,P(\mathbf{k},\,0)]^2}\qquad(24)$$

satisfies the condition

$$\widetilde{\rho}(k) < \rho(\mathbf{k}).\qquad(25)$$

The variance is expressed in terms of the spectral density

$$\sigma^2 = \int \rho(\mathbf{k})\,d\mathbf{k}.\qquad(26)$$

It follows that the screening always reduces the variance:

$$\widetilde{\sigma}^2 < \sigma^2.\qquad(27)$$

In the limit of an infinitely small lattice constant, the quantity $\rho(\mathbf{x})$ in these formulas remains proportional to the product of the concentrations, like $R(\mathbf{X})$ in Eq. (10), but σ may differ from σ_0 for the following reasons. The division into primitive cells, i.e., cells containing one atom of either component, is an arbitrary procedure which can be applied also to larger cells. The physical smallness of a cell implies the smallness of its dimensions compared with all the characteristic lengths of the problem, for example, the thermal wavelength of carriers, etc. Bearing in mind this point, we shall assume that the parameter σ differs from σ_0. Similarly, $R(\mathbf{X})$ is related to $\rho(\mathbf{x})$ only with an accuracy implied in the process of transition to infinitely small distances.

We can easily demonstrate the physical meaning of Eq. (27). A random potential represents a random distribution of potential wells and hillocks of various forms and dimensions. When carriers appear, they fill the wells, reduce the hillocks, and thereby level out the random function. The variance, which is a measure of the scatter of the potential with respect to its average value, is naturally reduced.

To obtain a more accurate estimate of the change in the variance due to screening, we shall use the fact that the parameter which has the dimensions of length in the proposed theory is the distance at which the quantity $w(\mathbf{x})$ vanishes; this quantity is the difference between the potentials associated with the atoms of the two components. We shall denote the distance in question by r_0. In non-Coulomb solid solutions, this distance is of the order of the interatomic spacing. However, in the general theory of the screening of static fields by free carriers, we have the following parameters with the dimensions of length: the thermal wavelength

$$\lambda = \frac{h}{(m\varkappa T)^{1/2}} \tag{28}$$

(where m is the effective mass, $\varkappa$ is Boltzmann's constant, and T is the absolute temperature); the Debye radius

$$r_D = \left(\frac{\varkappa T}{4\pi n e^2}\right)^{1/2} \tag{29}$$

(where e is the electron charge, divided by the permittivity of a crystal, and n is the carrier density); and the "quantum" radius

$$b = \left(\frac{h^2}{nme^2}\right)^{1/4}. \tag{30}$$

In the stationary random field approximation, we should regard the lattice constant as infinitely small and, consequently, we should assume that the parameter r_0 for non-Coulomb solutions is less than all the characteristic lengths (or, at least, equal to the shortest of these lengths). Accordingly, the "ultra-quantum" approximation, valid at short distances, should be used for the polarization operator:

$$P(\mathbf{k}, 0) = -\frac{mne^2}{(2\pi)^4 k^2} + O\left(\frac{1}{k^4}\right).$$

When these assumptions are made, we obtain

$$\widetilde{\rho}(\mathbf{k}) = \frac{\rho(\mathbf{k})}{[1 + (bk)^{-4}]^2}, \quad k > b^{-1}. \tag{31}$$

Thus, the screened correlation function $\widetilde{\rho}(\mathbf{k})$ is an order of magnitude smaller than $\rho(\mathbf{k})$ when $k \approx b^{-1}$.

The situation can be represented qualitatively as follows. The parameter r_0 represents the characteristic dimensions of the potential wells, due to the atoms of the components of the solution. If the wavelength of a carrier is less than the dimensions of a well, it is either captured by the well (this gives rise to local states), or the carrier wave function gives rise to more or less pronounced antinodes in the vicinity of the well. In either case, the charge is redistributed and the primary potential is smoothed out.

An effect of the same physical nature has been observed by Bonch-Bruevich and Glasko [15], who considered the behavior of hydrogen-like levels in a semiconductor, making allowance for the screening. They found that when the density of the screening carriers is sufficiently high, the local levels may be "pushed out" into an allowed band; this effect is not so much due to the Debye cutoff of the field at large distances as to the screening effect of carriers near a Coulomb center.

If there is some ionic binding in a solid solution, the correlation is known to vanish at distances greater than the Debye radius and the spectral density is given by

$$\widetilde{\rho}(k) = \frac{\rho(k)}{[1 - (kr_D)^{-2}]^2}, \ k^{-1} < r_D. \tag{32}$$

Thus, the smoothing out of the potential at short distances is stronger than that at long distances, but both effects act in parallel and reduce the variance.

Consequences which are finer than those represented by Eqs. (31) and (32) follow from the screening of the spectral density and apply to the differentiability of the random potential. By differentiability and integrability we mean these properties when applied to the rms values. It is known [12] that the existence of a derivative of a random function (for example, a derivative with respect to x_1) can be deduced from the integral

$$\int_{-\infty}^{\infty} k_1^2 \rho(k) \, dk, \tag{33}$$

which diverges for nondifferentiable functions and exists only for differentiable functions. The divergence may be due solely to the behavior of the spectral density at high values of k. It is clear that if the integral (33) exists for the primary spectral density, then it must exist for the screened spectral density. Moreover, in principle, we can have cases when the primary potential is not differentiable but the screened potential is. Such an effect may also be the consequence of the smoothing out of the potential by carriers.

Higher correlation functions can be treated similarly. In this case, the screening results in the appearance, in the denominators [as in Eqs. (23) and (24)], of expressions containing higher powers of the polarization operator. Thus, the screening can only improve the convergence of the expansion in terms of the correlation functions, and we may assume that there exists a range of temperatures in which the restriction to a pair correlation function is justified. Unfortunately, experimental data are needed to validate this theory, because general analysis is, for various reasons, still too difficult — more difficult, in fact, than in any variant of the perturbation theory in the absence of a random field.

In order to further justify the pair correlation function approximation, we shall consider only (as mentioned in the Introduction) those solid solutions which remain semiconductors at all concentrations in which they are capable of existing. Such solid solutions comprise the majority of semiconducting solid solutions, and the "metallization" of a solid solution due to a change in the concentration is a fairly rare phenomenon.

These results on the screening of stationary random fields by free carriers can be used not only in the theory of solid solutions, but also (in some cases, even more justifiably) in all cases when we know in advance that we are dealing with such random fields. This applies to glasses, polymers in bulk, and any "frozen" media, whose properties do not vary with time. This remark applies also to the subsequent parts of the treatment whenever we are speaking specifically of solid solutions.

3. MASS OPERATOR, CARRIER ENERGY SPECTRUM, AND DAMPING

We shall now see to what extent the presence of a screened stationary random field affects the dispersion law. For this purpose, it is convenient to use the method developed by Bonch-Bruevich and Mironov [16] for heavily doped semiconductors.

Let us assume that G is a retarded anticommutator Green's function for electrons in the absence of a random field. The equation for the complete Green's function G has the following symbolic form:

$$G_0^{-1}G + (\widetilde{\varphi} + \mathbf{M}\widetilde{\varphi})\,G = 1. \tag{34}$$

Here, $\widetilde{\varphi} + \mathbf{M}\widetilde{\varphi}$ is the total potential (we recall that in the preceding section we have considered a centered potential). Because of its stationary nature, the quantity $\mathbf{M}\widetilde{\varphi}$ is independent of the coordinates and, as shown in Section 1, is proportional to the concentration of one of the components; we shall assume that $\mathbf{M}\widetilde{\varphi} = n_1\Delta$, where Δ is a parameter with the dimensions of energy.

In principle, there are two possible cases of "linear interpolation." The first and simpler is that when there are no changes in the bands when one of the components is replaced completely with the other and electrons or holes are in the same energy minimum or maximum at all concentrations. In this case, the parameter n_1 varies from 0 to 1, and Δ represents the change in the forbidden bandwidth when one of the components is completely replaced with the other. Such a situation may (but need not) occur when the structures of the bands of the two components are basically the same; by this we mean the same number of critical points with the same sequence of these points on the energy scale. For example, we can see from Fig. 6 that the band structures of some $A^{III}B^V$ compounds are basically similar.

In the second case, the transfer of electrons (holes) to a given minimum (maximum) takes in only a certain range of concentrations (as in the case of Ge−Si solid solutions; cf. later part of the treatment), and then n_1 is restricted to that range of concentrations and Δ represents the change in the forbidden bandwidth when the concentration varies within this interval. Such a situation is always obtained when the band structures are basically different. We shall use the notation

$$G_0^{-1} + n_1\Delta = G_0'^{-1} \tag{35}$$

and rewrite Eq. (34) in the form

$$G_0'^{-1}G + \widetilde{\varphi}G = 1. \tag{36}$$

29

Multiplying this equation on the left by G_0^l, we obtain

$$G = G_0' - G_0' \widetilde{\varphi} G, \tag{37}$$

which gives the following expression after iteration:

$$G = G_0' - G_0' \widetilde{\varphi} G_0 + G_0' \widetilde{\varphi} G_0 \widetilde{\varphi} G. \tag{38}$$

The observed effects are described by average quantities. In particular, $\mathbf{M}G$ is the average Green's function. The existence of such a function must be proved; however, we shall assume, by way of hypothesis, that the average Green's function not only exists but has the same properties as the Green's function in the dynamic (not the probability) theory. If we make this assumption, we must assume also that the zeros of the function $(\mathbf{M}G)^{-1}$ determine the spectrum and damping of electrons. Physically, this hypothesis is equivalent to the assumption of the existence of some "average" quasi-particles, whose spectrum and damping are found using the same rules as those employed for the corresponding quantities in the dynamic theory.

Averaging of Eq. (38) gives

$$\mathbf{M}G = G_0 + \mathbf{M} \left(\widetilde{\varphi} \, (\mathbf{M}G) \, \widetilde{\varphi} \right) G_0 + \cdots . \tag{39}$$

The dots in the above expression represent terms of higher order in $\widetilde{\varphi}$, obtained in further iterations. We shall show later that the omission of these terms is equivalent to the omission of higher correlation functions. We shall define the mass operator $\mathfrak{M}$ by the relationship

$$(\mathbf{M}G)^{-1} = G_0^{-1} + \mathfrak{M} \tag{40}$$

or

$$\mathfrak{M} = - \mathbf{M} \{ \widetilde{\varphi} \, (\mathbf{M}G) \, \widetilde{\varphi} \}. \tag{41}$$

In the coordinate representation

$$\mathfrak{M} \, (x, y) = - \mathbf{M} \{ \widetilde{\varphi} \, (x) \, \widetilde{\varphi} \, (y) \} \mathbf{M} G \, (x, y) = - \widetilde{\rho} \, (x - y) \mathbf{M} G \, (x, y). \tag{42}$$

Here, x represents all the space and time coordinates, and y has the same meaning as x. Thus, the mass operator is, in this approximation, proportional to the correlation function. Higher corrections to the mass operator are related to higher correlation functions; they can be neglected because of the strong screening of the higher correlation functions. Since we shall deal with average quantities, it is convenient to represent $\mathbf{M}G(x, y)$ simply by $G(x, y)$ and to introduce a Fourier transform of the average Green's function

$$G \, (x, y) = \frac{1}{(2\pi)^4} \int d\mathbf{k} \, dk_0 e^{i \, (k_0 t + \mathbf{k} \, (\mathbf{x} - \mathbf{y}))} G \, (k, k_0), \tag{43}$$

so that

$$\mathfrak{M} \, (\mathbf{k}, k_0) = - \frac{1}{(2\pi)^3} \int dk' \widetilde{\rho} \, (\mathbf{k}') \, G \, (\mathbf{k} - \mathbf{k}', k_0). \tag{44}$$

In Eq. (43), it is assumed, in accordance with the hypothesis of stationary conditions, that $G(x, y)$ depends only on the difference $(\mathbf{x} - \mathbf{y})$; the difference between the corresponding times is denoted by t.

The possible values of the carrier energy are governed, as in the general theory [15], by the singularities of the corresponding spectral function $I(\mathbf{k}, k_0)$ in the lower half-plane (including the real axis). For real values of k_0, this spectral function is related to the imaginary part of the Green's function by the expression

$$I (\mathbf{k}, k_0) = \frac{2 \operatorname{Im} G (k, k_0)}{1 + e^{\beta (h k_0 - \mu)}}, \tag{45}$$

where $\beta^{-1} = \varkappa T$, and μ is the Fermi level. Thus, the equation for the energy spectrum of carriers can be obtained by equating to zero the denominator of the imaginary part of the Green's function and assuming that $k_0 = \omega' - i\omega''$ is complex. The Green's function of carriers in the absence of a random field has the form

$$G_0 (k, k_0) = - \frac{1}{(2\pi)^4} \frac{1}{h k_0 - \varepsilon (k) - n_1 \Delta + i\eta}, \quad \eta \to +0, \tag{46}$$

where $\varepsilon(\mathbf{k})$ is in this case given by Eq. (18). The origin for the energy is the bottom of the conduction band of one of the pure components (for $n_1 = 0$). At sufficiently high temperatures [cf. (27)], the variance of the screened potential is fairly small and the mass operator is also small. This allows us to assume that the damping is fairly small ($\omega'' \ll |\omega'|$). Using the definition of the mass operator, we find equations for the calculation of the spectrum and damping in the presence of a random field

$$h\omega' - \varepsilon (\mathbf{k}) - n_1 \Delta - \operatorname{Re} \mathfrak{M} (k, \omega') = 0,$$

$$h\omega'' = \frac{\lim\limits_{\omega'' \to 0} \operatorname{Im} \mathfrak{M} (\mathbf{k}, \omega' - i\omega'')}{1 - \dfrac{\partial \operatorname{Re} \mathfrak{M} (\mathbf{k}, \omega)}{\partial h\omega}}. \tag{47}$$

The problem is thus reduced to the finding of the mass operator, which can be calculated approximately by substituting into Eq. (44) the unperturbed Green's function (46):

$$\mathfrak{M} (\mathbf{k}, k_0) = - \frac{1}{(2\pi)^3} \int \frac{\widetilde{\rho} (\mathbf{k}') \, d\mathbf{k}'}{h k_0 - \varepsilon (\mathbf{k} - \mathbf{k}') - n_1 \Delta - \mathfrak{M} (\mathbf{k} - \mathbf{k}'_1 k_0)}. \tag{48}$$

The first iteration gives

$$\operatorname{Re} \mathfrak{M} (\mathbf{k}, \omega) \equiv \mathfrak{M}_1 = \frac{1}{(2\pi)^3} \int \frac{\widetilde{\rho} (\mathbf{k}') \, d\mathbf{k}'}{h\omega' - \varepsilon (\mathbf{k} - \mathbf{k}') - n_1 \Delta},$$

$$\lim_{\omega'' \to +0} \operatorname{Im} \mathfrak{M} (\mathbf{k}_1 \omega' - i\omega'') = \frac{\pi}{(2\pi)^3} \int \widetilde{\rho} (\mathbf{k}') \, \delta [h\omega' - \varepsilon (\mathbf{k} - \mathbf{k}') - n_1 \Delta - \mathfrak{M}_1 (\mathbf{k} - \mathbf{k}'_1 \omega')] \, d\mathbf{k}'. \tag{49}$$

In Eq. (49), we can replace $h\omega'$ by the unperturbed energy $\varepsilon(\mathbf{k}) + n_1\Delta$, thus excluding from our discussion a small region near $\mathbf{k} = 0$. From Eqs. (47) and (49), we find

$$h\omega' \approx \varepsilon (\mathbf{k}) + n_1 \Delta + \varepsilon_1 (\mathbf{k}).$$

The change in the dispersion law thus consists in a shift of the energy by an amount equal to the average potential and a correction which is a quadratic function of the concentrations of the components:

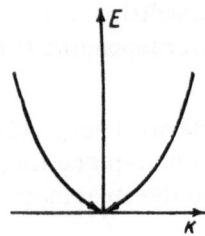

Fig. 1. Schematic representation of the dispersion law with the point **k** = 0 excluded.

$$\varepsilon_1(\mathbf{k}) = \frac{1}{(2\pi)^3} \int \frac{\widetilde{\rho}(\mathbf{k}') \, d\mathbf{k}'}{\varepsilon(\mathbf{k}) - \varepsilon(\mathbf{k}')} \cdot \tag{50}$$

This formula is inapplicable only near a minimum of $\varepsilon(\mathbf{k})$, because (in contrast to ideal crystals) the parabolic component of the dispersion law should be represented as shown in Fig. 1, excluding from the dispersion a small region around the point where the denominator in the integrand becomes small. However, as shown in the preceding section, $\widetilde{\rho}(\mathbf{k})$ tends to zero when $\beta \to 0$, and, consequently, the excluded region may become sufficiently small due to screening and the majority of electrons will thus be in the parabolic region (with a renormalized effective mass). Applying the theorem on means in order to take out the denominator outside the integral sign, we find that, in the thermal region,

$$\varepsilon_1(\mathbf{k}) = \frac{1}{(2\pi)^3} \frac{\widetilde{\sigma}^2}{\varepsilon(\mathbf{k}) - \varepsilon(\mathbf{k}_0)} \cdot \tag{51}$$

Thus, the smallness of the correction to the spectrum is due to the smallness of the variance compared with $\varkappa T$. We must stress that both $\widetilde{\sigma}$ as well as $\mathbf{k}_0$ have a nontrivial temperature dependence, and that such a dependence may indicate a considerable contribution of $\varepsilon_1(\mathbf{k})$ to the observed dispersion law. Using Eq. (51), we can give a clear interpretation of the correlation function. To do this, we note that $\varepsilon_1(\mathbf{k})$ is simply a correction calculated from the perturbation theory, and that $\widetilde{\rho}(\mathbf{k})$ is a quantity proportional to the scattering cross section when the scattering is regarded as a perturbation. If $\widetilde{\rho}(\mathbf{k})$ decreases monotonically along a fixed direction there are no resonance effects (of the Bragg reflection type) in the scattering or, in other words, the random potential does not contain periodic components, so that the electronic properties of a solid solution differ from the corresponding properties of a continuous medium only by the renormalization of the parameters (the effective mass, etc.). If the order in the atomic distribution affects the electronic properties of a solution, such an effect is manifested by oscillations of $\widetilde{\rho}(\mathbf{k})$.

4. CURVED BAND APPROXIMATION

As mentioned before, in the effective mass approximation, the "primary" field may not be differentiable but, because of the screening, it may become differentiable at sufficiently high temperatures. At not-too-high temperatures, the smoothing out of the primary field may result in small derivatives of the potential, rather than in a small potential. The potential should become small at lower temperatures than those at which the derivatives of the potential are small, because the smoothness of the function usually means that the higher derivatives are small compared with the lower ones. Formally, this can be demonstrated as follows. It is known that if $\widetilde{\varphi}(\mathbf{x})$ is a differentiable stationary random function associated with a correlation function $\widetilde{\rho}(\mathbf{x})$, then the correlation function of the derivative of $\widetilde{\varphi}(\mathbf{x})$ with respect to x_i is $\partial^2\widetilde{\rho}/\partial x_i^2$. Thus, an expansion of the potential, for example, as a Taylor's series, corresponds to an expansion of the correlation function as a series in terms of the derivatives. Applying the operator ∇^2 to $\widetilde{\rho}(\mathbf{x})$, we find that the smallness of the correlation function of the potential gradient, particularly the smallness of its variance, is ensured by the smallness of the integral

$$\int \mathbf{k}^2 \widetilde{\rho}(\mathbf{k}) e^{i\mathbf{kx}} \, d\mathbf{k}.$$

This condition is less stringent than the condition of the smallness of $\widetilde{\sigma}^2$, because the screened spectral density $\widetilde{\rho}(\mathbf{k})$ vanishes sufficiently rapidly at large values of $|\mathbf{k}|$, even when it is large at low values of $|\mathbf{k}|$. We shall describe a variant of the perturbation theory suitable for small gradients of the function $\widetilde{\varphi}(\mathbf{x})$ without special restrictions as to the value of the variance. This means that we shall deal with solid solutions whose components have long-range – for example, Coulomb – potentials.

We shall start, like Bonch-Bruevich [17, 17a], from an equation for the retarded one-fermion Green's function G, which can be expressed symbolically thus:

$$(\omega + C - T - \widetilde{\varphi} - \mathfrak{M})\,G = -1. \tag{52}$$

Here, T is the kinetic energy operator; $\omega = E + \mu$ is the energy measured from the bottom of the conduction band; C is the renormalization constant, in which we shall later include the constant which centers the random function $\widetilde{\varphi}$; and $\mathfrak{M}$ is the mass operator. In an expanded form, we obtain two associated equations:

$$\left\{ E + \mu + C + i\eta + \frac{h^2}{2m}\,\Delta\mathbf{x} - \widetilde{\varphi}(\mathbf{x}) \right\} G(\mathbf{x}, \mathbf{y}; E) - \int dz \mathfrak{M}(\mathbf{x}, \mathbf{z}; E)\,G(\mathbf{z}, \mathbf{y}; E) = -\delta(\mathbf{x} - \mathbf{y}), \tag{53}$$

33

$$\left\{ E + \mu + C + i\eta + \frac{h^2}{2m} \Delta \mathbf{y} - \widetilde{\varphi}(\mathbf{y}) \right\} G(\mathbf{x}, \mathbf{y}; E) - \int d\mathbf{z}\, \mathfrak{M}(\mathbf{z}, \mathbf{y}; E) = -\delta(\mathbf{x} - \mathbf{y}). \tag{54}$$

The mass operator allows also for the interaction between electrons. We can show [17] that at not-too-high temperatures such an interaction may be allowed for by replacing the kinetic energy operator with a more complex expression; in the simplest case, this can be done by the renormalization of the effective mass. We shall consider only this simple case, and we shall assume also that the kinetic energy operator is renormalized also to allow for the "truncating" effect of the screening described earlier. It follows that m in Eqs. (53) and (54) becomes a function of the concentration of the components of a solid solution and of temperature. Bearing this in mind, we add Eqs. (53) and (54) and divide the resultant sum by 2; we also subtract these two equations, thus obtaining

$$\left\{ \omega + C + i\eta + \frac{h^2}{2m} \nabla_R^2 - \widetilde{\varphi}(\mathbf{R}) - \frac{1}{8} r_\alpha r_\beta \frac{\partial^2 \widetilde{\varphi}}{\partial R_\alpha \partial R_\beta} \right\} G(\mathbf{r}, \mathbf{R}; E) = -\delta(\mathbf{r}),$$

$$\left\{ \frac{h^2}{m} (\nabla_r \nabla_R) - r_\alpha \frac{\partial \widetilde{\varphi}(\mathbf{R})}{\partial R_\alpha} \right\} = 0. \tag{55}$$

Here, we have introduced the relative coordinates

$$\mathbf{x} + \mathbf{y} = 2\mathbf{R}, \quad \mathbf{x} - \mathbf{y} = \mathbf{r}$$

and we have dropped the higher derivatives of the potential because they are small. Double coordinate indices represent summation. The second expression in Eq. (55) can be regarded as an additional requirement, which the solution of the main equation must satisfy. It is convenient to introduce dimensionless variables by measuring the energy in some characteristic units, for example, in units of $\widetilde{\sigma}$ or $\varkappa T$, and the length in units of $h^2/(2m\widetilde{\sigma})^{1/2}$ or $h/(2m\varkappa T)^{1/2}$. Then the symbolic solution of Eq. (55) can be written in the form

$$G(\mathbf{r}, \mathbf{R}; E) = i \int\limits_0^\infty ds\, e^{-\eta s + i\omega s + iCs} L\delta(\mathbf{r}), \tag{56}$$

where

$$L = \exp is \left\{ \nabla_r^2 + \frac{1}{4} \nabla_R^2 - \widetilde{\varphi}(\mathbf{R}) - \frac{1}{8} r_\alpha r_\beta \frac{\partial^2 \widetilde{\varphi}(\mathbf{R})}{\partial R_\alpha \partial R_\beta} \right\}. \tag{57}$$

We shall use the smallness of the derivatives once again and "extract" approximately the operator factors $\exp\left(\frac{is}{4} \nabla_R^2\right)$ and $\exp(is\nabla_r^2)$ from the exponential function in Eq. (57). For this purpose we can use the following property of the operators of the type [17a]

$$L = \exp[\sigma(U + \eta T)],$$

where u and T are linear operators, σ is a continuous parameter, and η is a pure number. We shall assume that

$$L = S \exp(\sigma\eta T).$$

Then, the following equation applies to the operator S:

$$\frac{\partial S}{\partial a} = US + \eta[T, S],$$

where $[T, S] = TS - ST$ and the "initial condition" is $S(0) = 1$. However,

$$[T, S] = e^{\sigma(U + \eta T)} \sum_{l \geq 1} \frac{\sigma^l}{l!} K_l e^{-\sigma \eta T}.$$

Here, the K_l are calculated from the recurrent formula

$$K_l = [K_l - 1, U + \eta T], \qquad K_1 = [T, U].$$

In the case of interest to us, $T = -\nabla_R^2$ and $U = \widetilde{\varphi}(R)$. Using Fourier transforms, we obtain

$$G(r, R; E) = i \int_0^\infty ds \int d\mathbf{k} \exp \{- \eta s + is [\omega + C - \mathbf{k}^2 - \widetilde{\varphi}(R)] +$$

$$+ i\mathbf{kr} + \frac{is^2}{4} \nabla_R^2 \widetilde{\varphi}(R) - \frac{is^3}{12} (\nabla_R \widetilde{\varphi}(R))^2 - \frac{is^3}{6} k_\alpha k_\beta \frac{\partial^2 \widetilde{\varphi}(R)}{\partial R_\alpha \partial R_\beta} + \frac{is^2}{2} r_\alpha r_\beta \frac{\partial^2 \widetilde{\varphi}(R)}{\partial R_\alpha \partial R_\beta} - \frac{is}{8} r_\alpha r_\beta \frac{\partial^2 \widetilde{\varphi}(R)}{\partial R_\alpha \partial R_\beta} \}. \qquad (58)$$

The average Green's function $MG(\mathbf{r}, \mathbf{R}; E)$ in this stationary case should be calculated using a distribution function independent of $\mathbf{R}$ and its dependence on r should be given by the form of the correlation function. It follows from Eq. (58) that if all the derivatives of the potential are dropped, the Fourier transform $G(\mathbf{k}, E)$ of the average Green's function, given by the equation

$$MG(\mathbf{r}, \mathbf{R}; E) = \int d\mathbf{k} e^{i\mathbf{kr}} G(\mathbf{k}, E),$$

should be calculated regarding $\widetilde{\varphi}(R)$ simply as a random quantity and not a random function (or, if it is convenient, a random function with independent values). Such an approximation corresponds to the curved band approximation, i.e., the assumption that at each point we have some fixed bandwidth governed by the value of the random potential at that point, and that this width changes very little at distances comparable with the wavelength of carriers. In transport problems, the curved band approximation is not always suitable; however, in thermodynamic problems, which will be the only ones considered here, we can use it, bearing in mind the limitations stated here.

We shall calculate

$$G(\mathbf{k}, E) = i\mathbf{M} \int_0^\infty ds e^{s \{-\eta + i (\omega + C - \mathbf{k}^2 - \widetilde{\varphi})\}}. \qquad (59)$$

Introducing a characteristic function of the quantity $\widetilde{\varphi}$

$$f(s) = \mathbf{M} e^{is\widetilde{\varphi}},$$

we find that

$$G(\mathbf{k}, E) = i \int_0^\infty ds f(-s) e^{s [-\eta + i (\omega + C - \mathbf{k}^2)]}. \qquad (60)$$

We shall define the cumulant function $\psi(s) = \ln f(s)$ in the usual manner. Then

$$G(\mathbf{k}, E) = i \int_0^\infty ds\, e^{\psi(-s) + s\,[-\eta + i\,(\omega + C - \mathbf{k}^2)]}. \tag{61}$$

Let us assume that the random quantity $\widetilde{\varphi}$ has an absolute moment of the p-th order, so that the characteristic function can be differentiated p times. Expanding the cumulant function as a series of semi-invariants, we find that

$$G(\mathbf{k}, E) = i \int_0^\infty ds\, \exp s \left\{ -\eta + i\,(\omega + C - \mathbf{k}^2) - \sum_{l=1}^p i^{-l} \frac{\gamma_l}{l!} (-s)^{l-1} + \frac{1}{s} O(s) \right\}. \tag{62}$$

Here,

$$\gamma_l = i^{-l} \left(\frac{d^l \psi(s)}{ds^l} \right)_{s=0}$$

is a semi-invariant of the order of l and $O(s)$ is the nondifferentiable part of the cumulant function. For the quantity μ to be equal to the true chemical potential, we must assume that $C = \gamma_1$. If we bear in mind that a first-order semi-invariant is equal to the mathematical expectation (multiplied by i), this result becomes easy to understand in the physical sense: when we speak of the average Green's function, we should shift the origin of the energy by the average value of the random potential.

These formulas are convenient for the calculation of the thermodynamic properties of a system of fermions from the equation

$$n = \int_{-\infty}^\infty \frac{D(E)\, dE}{e^{\beta E} + 1}, \tag{63}$$

where n is the number of particles,

$$D(E) = \frac{1}{4\pi^2} \int d\mathbf{k}\, \operatorname{Im} G(\mathbf{k}, E). \tag{64}$$

Formulas (60)-(64) give a complete general solution of the problem of the thermodynamic properties of a gas of quasi-classical fermions in the presence of external random fields independent of time. It also follows from these formulas that, for a system to behave in a random field in the same way as in the absence of fields, the function $\psi(-s) + iCs$ must, for some values of C, depend sufficiently weakly on s.

The semi-invariants can be expressed rationally in terms of moments. Therefore, the foregoing formulas give the properties of the system not only when the random quantity is given in terms of semi-invariants, but also when it is represented by moments. When the distribution function $F(\widetilde{\varphi})$ and the characteristic function are given, it is convenient to use the relationship

$$D(E) = \frac{1}{2\pi^2} \int_0^\infty x^{1/2} F'(x - \omega)\, dx, \tag{65}$$

which can be easily obtained from Eqs. (60) and (64), from the property $f(s) = \overline{f}(-s)$ of the characteristic function (the bar represents complex conjugation), and from the inversion formula

$$F'(\widetilde{\varphi}) = \frac{1}{2\pi} \int_{-\infty}^{\infty} e^{-is\widetilde{\varphi}} f(s)\, ds,$$

where $F'(\widetilde{\varphi})$ is the probability density for the quantity $\widetilde{\varphi}$. Equation (65) has a simple meaning: it can be regarded as the convolution of the density of states in an ideal Fermi gas in the absence of a random field and the distribution $F(\widetilde{\varphi})$, so that the fermion energy is equal to the sum of the energy in the absence of a field and the random potential $\widetilde{\varphi}$. However, Eq. (65) must not be given the literal meaning of the convolution, since the density of states in an ideal Fermi gas

$$D(E) = \frac{1}{2\pi^2}\, \omega^{1/2},\ \ \omega > 0,$$
$$D(E) = 0,\ \ \omega < 0 \tag{66}$$

does not have even the formal features of the probability density distribution, because it is not normalized to unity. The reason for this is not physical, at least in the problems associated with the behavior of electrons in crystals and in condensed media. In fact, the function (66), used in the convolution (65) as a probability density distribution, has, in fact, a different form: after a more or less extended region where the dependence of the energy on the momentum is quadratic, the function passes through a maximum and decreases.

The ratio of the variance to the width of the forbidden band is one of the dimensionless parameters which determine the limits of applicability of the theory. The other energy parameter is the energy at which the nonparabolicity of the dispersion law becomes unimportant, i.e., the deviation from Eq. (66) becomes negligible.

Equation (65) can be obtained also from very simple considerations by assuming an ensemble of ideal Fermi gases, each of which is in some constant field, and averaging over all such fields with a given distribution $F(\widetilde{\varphi})$.

Using Eq. (65), we can calculate the density of states when the distribution function is given. The simplest case is a degenerate distribution. The random quantity then ceases to be random in its intrinsic sense if its distribution function becomes a function of a discontinuity (and the probability density becomes correspondingly a δ function): $F'(\widetilde{\varphi}) = \delta(\varphi_0 - \widetilde{\varphi})$, where φ_0 is the potential of a constant field. In this case, the density of states has the form given by Eq. (66), which is as expected.

It is worth noting that in a constant field the density of states is the same as in the absence of such a field. Otherwise, it would depend on the parameters of the distribution, i.e., on the spectral composition of the external field. This conclusion can be utilized also if we have to alter the density of states in some definite manner from, for example, the Maxwellian density distribution, considering Eq. (65) as an equation for $F'(\widetilde{\varphi})$, where $D(E)$ is known. For this purpose, we can also use fields which depend on time if the characteristic frequencies of these fields are considerably higher than the reciprocals of all the relaxation times, so that the thermodynamic approach is justified.

Irrespective of the nature of the distribution function, the cumulant function can be expanded as a series in terms of semi-invariants; in this case, for the reasons stated earlier, the variance cannot become infinite at sufficiently high temperatures. Therefore, when a second semi-invariant is used, we can regard the distribution function as normal.

Calculation of the integral (65) with a normal probability distribution

$$F'(\widetilde{\varphi}) = \frac{1}{\sigma\sqrt{2\pi}} \exp\left(-\frac{\widetilde{\varphi}^2}{2\widetilde{\sigma}^2}\right)$$

gives the expression

$$D(E) = \frac{\widetilde{\sigma}^{1/2}}{4\pi^2\sqrt{2}} e^{-\frac{\omega^2}{4\widetilde{\sigma}^2}} D_{-3/2}\left(-\frac{\omega}{\widetilde{\sigma}}\right), \tag{67}$$

where $D_p(x)$ is a parabolic cylindrical function of the order $-\frac{3}{2}$ (a Weber function).

When the variance tends to zero, the normal distribution tends to a degenerate state. Using asymptotic estimates of the function $D_p(x)$, we can show that the density of states then tends to Eq. (66), which is as expected. If, therefore, $\widetilde{\sigma} \ll \varkappa T$, the thermodynamic (but not, generally speaking, the transport) properties of the Fermi gas in a random field are practically identical with its properties in the absence of such a field.

We shall now consider the case of arbitrary variance, high temperatures, and Maxwellian electrons, when Eq. (63) becomes

$$n = \frac{\widetilde{\sigma}^{3/2}}{4\pi^2\sqrt{2}} e^{\beta\mu} \int_{-\infty}^{\infty} dx\, D_{-3/2}(-x)\, e^{-\frac{x^2}{4} - \widetilde{\sigma}\beta x}.$$

The integral on the right-hand side of this expression can be calculated using the following integral representation of the function $D_p(z)$ for negative values of p:

$$D_p(z) = \frac{e^{-\frac{z^2}{4}}}{\Gamma(-p)} \int_0^{\infty} e^{-zt - \frac{t^2}{2}} t^{-p-1} dt.$$

We find that

$$n = \frac{1}{8\pi^2} e^{\beta\mu} (\varkappa T)^{3/2} e^{\widetilde{\sigma}^2/2\,(\varkappa T)^2}.$$

This expression differs only by the factor $e^{\widetilde{\sigma}^2/2(\varkappa T)^2}$ from the corresponding expression for a Maxwellian gas in the absence of a random field. It follows from it that, irrespective of the value of the variance, at sufficiently high temperatures in the presence of a random field, electrons behave as if they were in an ideal crystalline semiconductor. Deviations from the usual temperature dependences of the thermodynamic properties of electrons in semiconductors should be observed at $\widetilde{\sigma} \approx \varkappa T$. This absence of a basic difference between the behavior of electrons in solid solutions and in ideal crystals indicates that the selected approximation of two first semi-invariants is sufficient to describe its thermodynamic properties in the media considered.

It follows from this conclusion that the presence of a random field is manifested by a "tail" in the density of states, which is observed, in particular, in the determination of the forbidden bandwidth by optical methods: it appears in the form of a broadened edge of the absorption band. We must remember that the "tail" observed in these experiments is due to the broadening of the bottom of the conduction band as well as the top of the valence band. In the case of one band, the density of states for small values of the variance and (or) far from the former edge of the band has the form

$$D(E) = D^0(E) + \frac{\widetilde{\sigma}^2}{4\pi^2 \sqrt{2}} e^{-\omega^2/2\widetilde{\sigma}^2} |\omega|^{-3/2}, \quad \omega \neq 0,$$

where $D^0(E)$ is given by Eq. (66). This formula is obtained from Eq. (67) and from the asymptotic form of the parabolic cylindrical functions

$$D_p(z) \approx e^{-z^2/4} z^p \left[1 + 0\left(\frac{1}{z^2}\right) \right], \quad |z| \gg 1; \quad |z| \gg p.$$

In addition to normal distributions, we can consider also other distributions without expanding the cumulant function in terms of semi-invariants. One of the distributions which can be given a simple physical interpretation is the Cauchy distribution with a characteristic function

$$f(s) = e^{ias - b|s|},$$

where a and b are parameters. The parameter a is always included in the renormalization constant, and b can be identified with η in Eqs. (60)-(62), without assuming that it approaches zero. Thus, if the value of σ has the Cauchy distribution, the presence of a random field is manifested by finite damping and by the renormalization of the effective mass, i.e., its dependence on temperature, on the concentration of the components, and on the carrier density.

The normal distribution and the Cauchy distribution belong to a class of stable distributions, which can be defined as follows [27]. Let us assume that ξ is a random quantity (in our case, the potential φ at a given point) and $F(\xi)$ is its distribution function, which has the following property: for any real numbers $a_1 > 0$, $a_2 > 0$, b_1 and b_2, there are such numbers a and b that $F(a_1\xi + b_1) * F(a_2\xi + b_2) = F(a\xi + b)$, where the asterisk represents the convolution; then $F(\xi)$ is known as a stable distribution function. We recall that the convolution of the two distributions, $F_1(\xi)$ and $F_2(\xi)$, is the distribution function of the sum $\xi + \xi'$.

Thus, the stability is a group-theoretical concept: the distribution function should be invariant, in the sense described here, with respect to a linear transformation of a random quantity followed by addition. We shall return to stable distributions in the discussion of transport problems.

The use of various distribution functions as mathematical models does not give accurate information until we have a sufficient amount of experimental data to justify a given model. At present, such experimental data are lacking; therefore, in addition to such models many useful empirical rules have been obtained, which are suitable in qualitative interpretation. For example, the following rule [196] makes it possible to select a solid solution with the required properties: the doping of a semiconductor compound with another compound having a larger ratio R_M/R_N, where R_N and R_M are, respectively, the ionic and metallic radii, reduces the ratio of the hole mobility to the electron mobility.

In the next section, we shall describe a model in which the main empirical assumption is that the concepts developed in the theory of ideal crystals apply also to solid solutions.

5. VIRTUAL CRYSTAL MODEL

In theoretical treatments, dealing with various properties of solid solutions, it is assumed that the potential in Schrödinger's equation (or, if it is more convenient, in the effective wave equation) consists of two components: a periodic component with the period of the Bravais lattice of an "averaged" solid solution (virtual crystal) and the nonperiodic component due to the random distribution of atoms:

$$\widetilde{\varphi}(\mathbf{x}) = \widetilde{\varphi}_1(\mathbf{x}) + \widetilde{\varphi}_2(\mathbf{x}). \tag{68}$$

The nonperiodic component $\widetilde{\varphi}_2(\mathbf{x})$ is regarded as a perturbation. First of all, this approach is equivalent to the assumption that the variance is small, and, consequently, the correlation function is small for any value of the argument. In fact, when the variance approaches zero, the probability density approaches the δ-type form, and the distribution function is that of an ideal crystal in which the potential is equal to the average value of the distribution function. Secondly, if we write the potential in the form of Eq. (68) without any special assumptions about the nature of the nonperiodic component of the potential, we imply that the correlation function vanishes at distances shorter than all the characteristic lengths (in particular, the thermal wavelength of electrons). In view of this, the a priori assumption about the possibility of separating Eq. (68) into two components requires some refinement. The point is that the periodic component of the potential may apply to a unit cell or to any group of cells, geometrically similar to a unit cell and consisting of a very large but finite number of such cells: when we deal with an infinite crystal, the division into large cells gives the same geometrical distribution. If an ideal infinite crystal has physical properties which are invariant under the translation of the whole crystal by the lattice vector, then an average (i.e., real) solid solution has similar properties only when it is translated by such a lattice vector whose length is longer than all the characteristic lengths and, in particular, the lengths at which the correlation function of the potential vanishes. The introduction of the correlation function avoids the difficulty associated with the ambiguity of the division of Eq. (68) into two parts and reduces the problem to a comparison of different characteristic lengths, having simple physical and probability meanings. Direct analysis is given in several papers [155] of the various "higher harmonics," which are obtained when various "superlattices" are constructed, and of the associated long-range order effects. (The quotes are used to indicate that we are not speaking of higher harmonics in the sense of Fourier expansions, of which there may be an arbitrary number in the potential for a unit cell, but that we mean the periodicity in the case of translation by vectors which are multiples of the fundamental lattice vectors.) In our case, the possible long-range order effects are contained, at least parametrically, in possible oscillations of the spectral density, since such oscillations can be considered to be an indication of a predominantly "resonant" scattering of electrons with certain definite wavelengths.

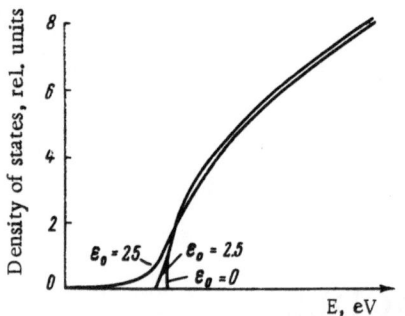

Fig. 2. Density of states obtained in the virtual crystal model [155].

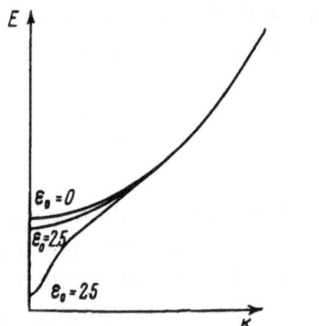

Fig. 3. Dispersion law in the virtual crystal model [155].

Direct calculations within the framework of the virtual crystal model have been carried out by Muto [179a] and Parmenter [155, 184-186]. Figure 2 shows the densities of states obtained by Parmenter, where ε_0 represents the energy at which the density of states vanishes. Figure 3 shows the dispersion laws found for different parameters of the problem. Analysis of the low-energy part of the dispersion law has led Parmenter to the conclusion that the transition from an ideal crystal to a solid solution should reduce the effective masses. The corresponding increase in the mobility should be compensated by the scattering on the random part of the potential. In terms of our treatment, it means that a quadratic dispersion law (cf. Fig. 1) cannot be extrapolated to zero.

Using the virtual crystal model, Parmenter deduced another effect — reduction in the anisotropy of the effective mass when the concentration of one component is increased. In fact, this effect is due not so much to the random field as to the self-consistent nature of this field, because of which it becomes isotropic.

The perturbation theory, presented in the preceding sections, differs basically from the virtual crystal model. This is because, in the virtual crystal model, it is assumed that the problem in the zeroth approximation is reduced, in some way, to the one-electron problem, which is usually taken to be independent of temperature, even in the parametric sense. The perturbation is considered within this one-electron framework. In the improved perturbation theory, a variant of which is presented in Sections 2-5, the screened field is regarded as the perturbation; the "primary" field can be arbitrary and, thus, the unperturbed and perturbed problems are many-particle problems, at least in the sense of a parametric dependence on temperature.

The models described in this and preceding sections make it possible to deal with typical solid solutions. For example, if we are dealing with a solid solution in which a change of a component does not change the ionic contribution to the potential, it is reasonable to use the renormalization procedure described in Section 4, or even the virtual crystal model; in other cases, a considerable number of effects due to long-range interactions can be allowed for also in the approximation described in Section 5. However, we must bear in mind that in the case of a considerable ionic binding, polaron effects may complicate the problem considerably. The representation of a quasi-particle as an entity possessing crystal momentum is usually retained and only the quantitative characteristics of the problem are altered. This makes it possible to apply directly the theory of ideal crystals to some solid solutions, replacing all the quantities with their renormalized values. Therefore, we shall give only a qualitative description of the observed effects, stressing where necessary the special properties of solid solutions.

6. OPTICAL PROPERTIES AND TRANSPORT PHENOMENA

It follows from the preceding sections that the behavior of carriers in solid solutions should differ greatly from their behavior in an ideal crystal only at a sufficiently low temperature, when the random field is not smoothed out enough by the carrier screening. Such temperatures may, in fact, be fairly high so that the random field may affect measured quantities. The optical electronic properties of ideal crystals depend both on the structure of the electron energy bands and on the phonon properties of a crystal. Before analyzing the additional influence of departures from ideal conditions, typical of a solid solution, we shall consider qualitatively the possible influence of such departures on the phonon spectrum.

In the simplest approximation a solid solution is replaced with some virtual crystal in accordance with the procedure described earlier. Herman, Glicksman, and Parmenter [155] have shown that in this approximation one can describe satisfactorily the low-energy parts of the acoustical modes, so that the phonon processes in a solid solution can be analyzed in the same way as in an ideal crystal if we can be sure that only the low-energy acoustical phonons participate in the phenomena considered. A rigorous treatment of longitudinal phonons in a one-dimensional chain has been given by Dyson [29]. As expected, the final results depend considerably on the parameters of the distribution used to carry out the averaging. Rezanov and Masharov [80a] have investigated the vibrational properties of substitutional solid solutions at such low impurity concentrations that the influence of disorder can be neglected. The structure of the vibrational spectrum, determined by the method of variation of the natural frequencies, has been found to be the same as that for the lattice of a pure solvent, but natural frequencies are now replaced by some renormalized frequencies depending on the concentration. A calculation of the lattice component of the thermal conductivity using the virtual crystal model in the long-wavelength approximation has shown, in agreement with numerous measurements, that the dependence of the thermal conductivity on the composition of a binary solution is determined by the product of the concentrations because the thermal resistance has a maximum at approximately equal concentrations of the components.

The general situation for phonons is similar to that for electrons, which have been discussed earlier. At low energies or low concentrations and not-too-high energies, phonons behave basically in the same way as in an ideal crystal, except that the measured quantities — for example, the velocity of sound — depend on the concentration of the components. The temperature dependence is weaker for phonons and, therefore, the problem of vibrations can be dealt with in the harmonic approximation by an exact separation of variables; and the effect on phonons due to interactions is not as strong as in the case of the Coulomb interaction of electrons. It follows that if the electron band structure is known correctly, interband transitions can be regarded qualitatively in the same way as in the theory of ideal crystals. Various concentration dependences, characteristic of solid solutions only, may appear in the line profile or in similar details. Analysis of interband transitions in Ge−Si and other solutions (see later part of this monograph) makes it possible to give a general description of the dependence of the band structure on the concentration.

Optical transitions involving impurity states should be more sensitive to the presence of a random field simply because of the comparatively low energies of these transitions. In particular, a shallow impurity state may disappear altogether in the "tail" of the density of states. One of the possible explanations of the absence of impurity conduction in glasses is based on an assumption that this happens in the vitreous state. In any case, in the presence of a random field, the "pushing out" of an impurity level into the conduction band should take place at a lower temperature than in the absence of a random field. Analysis of shallow local traps within the framework of the hydrogen-like model is, in this case, justified as much as in the case of ideal crystals, because the Bohr orbit radius in typical cases represents several hundreds or thousands of interatomic spacings; one has to regard the quantities obtained as the averages. Experimental investigations of such average quantities is one of the methods for determining the parameters of random fields acting in solid solutions.

When we consider external influences on a system of carriers, we must distinguish mechanical and thermal perturbations. The former are those which can be included explicitly in the Hamiltonian of the system; they include spatially uniform random fields, which are of interest to us, as well as external electric and magnetic fields. The thermal perturbations include, for example, temperature gradients or other quantities, which are thermodynamic rather than mechanical. For perturbations of the former type, Kubo [71] obtained general formulas for the transport coefficients in terms of time correlation functions. Zvyagin [51] considered the problem of calculating transport coefficients which describe the reaction of electrons in heavily doped semiconductors to mechanical and thermal perturbations. The random field acting on a carrier in such a semiconductor was assumed, as in the preceding case, to be proportional to a small parameter and a suitable perturbation theory was developed. The problem was solved in two ways. First, using Kubo's theory, the average density of the current and the average density of the energy flux were calculated in the lowest approximation with respect to the random field; the results were formally the same as those obtained in the treatment of these phenomena by the transport equation method. The only difference was that all the quantities were replaced by their average values. Secondly, the same problem was considered by solving directly the equation for the density matrix and the same result was obtained. Higher corrections to the transport coefficients were also found.

From these general considerations, it is clear that in problems of this type we may meet two mechanisms of carrier scattering: the "dynamic" reflection from each local departure from uniformity, and the "diffuse" scattering of electrons from many inhomogeneities at the same time. The situation can be represented qualitatively by considering the scattering of light by a turbid medium, when a ray of light retains some of its properties, while its other properties (for example, the velocity of propagation) change and at the same time the light is scattered diffusely by local inhomogeneities. In the exact treatment of the dynamic problem, all the scattering events are naturally dynamic and the diffusion component appears when we go over from the dynamic to the statistical description. Zvyagin allowed only for the "dynamic" interactions of electrons with impurity atoms and ignored the effects which would have resulted in an equation of the Fokker−Planck type rather than the Boltzmann equation. Formally, the omission of the diffusion effects is represented by the use of the theorem on averaging over an ensemble, due to Kohn and Luttinger [70]. This theorem is one of the variants of the limit theorems, usually obtained by the method of characteristic functions for sums of independently distributed terms: it is valid at not-too-high impurity concentrations.

We shall give a semiphenomenological derivation of the expression for the electrical conductivity, based on a parametric allowance for the dynamic and diffusion properties of the medium and the introduction of a random field of currents.

The general theory [15] shows that to calculate the electrical conductivity, it is suffi-cient to know the function

$$\langle [j_i(x), j_l(x')]_+ \rangle = F_{il}(x, x'),$$ (69)

where the angular brackets represent thermodynamic averaging, $j_l(x)$ is the l-th component of the density of the current at a point $x = \{\mathbf{x}, x_0\}$ in space−time. We shall assume that a sys-tem of electrons is in a random field, which produces a random field of currents $j_l(x)$. The probability characteristics of this random field of currents will be assumed to be known, and the problem is to express the electrical conductivity in terms of these characteristics.

The function $\mathbf{M}F_{il}(x, x')$, where the symbol $\mathbf{M}$ represents averaging over all space−time realizations of external fields, has a spectral representation (h = 1)

$$\mathbf{M}F_{il}(x, x') = \int\limits_{-\infty}^{\infty} dE \int d\mathbf{k} e^{-iEt-i(\mathbf{k}, \mathbf{x}-\mathbf{x}')} F_{il}(\mathbf{k}, E),$$ (70)

where $F_{il}(\mathbf{k}, E)$ is the spectral density. The possibility of such a representation follows from the fact that the field of currents is stationary; we shall consider only the stationary case. This limitation, together with the Hermitian nature of the current density operator, results in a real positive definite function $F_{il}(\mathbf{k}, E)$ [15]. The difference from the dynamic problem lies in the fact that a process with a correlation function $F_{il}(\mathbf{k}, E)$ has singularities, which depend on the nature of the random field of currents.

To avoid misunderstanding, we must stress that $F_{il}(\mathbf{k}, E)$ is the Fourier transform of the correlation function, obtained by double Gibbs averaging over all realizations of the ex-ternal random field; it differs from the Fourier transform of the correlation function which is obtained only by thermodynamic averaging. From now on we shall assume, for the sake of simplicity, that $F_{il}(\mathbf{k}, E)$ is the spectral density of a uniform and isotropic (with respect to $\mathbf{x}$) field.

We shall consider a general formula for the electrical conductivity [15]:

$$\sigma_{il}(\omega) = \frac{ine^2}{m\omega} \delta_{il} - \frac{i}{\omega} \int\limits_{-\infty}^{\infty} dE \frac{\tanh\frac{\beta E}{2}}{E - \omega - i\eta} \int d\mathbf{x} F_{il}(\mathbf{x}, \mathbf{x}'; E),$$ (71)

where n is the density of particles; e is the electron charge; m is some average effective mass, which may depend on temperature; and ω is the frequency. The left-hand and right-hand parts of Eq. (71) are functionals of the random field: the average (observed) value of the electrical conductivity is obtained by averaging Eq. (71). Because of the stationary and isotropic spatial conditions, we have

$$\mathbf{M}\sigma_{il}(\omega) = \sigma(\omega)\delta_{il},$$

$$\sigma(\omega) = \frac{ine^2}{m\omega} - \frac{i}{\omega} \int dE \frac{\tanh\frac{\beta E}{2}}{E - \omega - i\eta} F(0, E).$$ (72)

If F(k, E) is analytic with respect to E in the upper half-plane, it follows that the random field does not absorb energy, and

$$\sigma(\omega) = \frac{ine^2}{m\omega} - \frac{i}{\omega}(2\pi)^3 \int_{-\infty}^{\infty} dE \mathscr{P} \frac{\tanh\frac{\beta E}{2}}{E-\omega} F(0, E) + (2\pi)^3 \frac{\pi \tanh\frac{\beta E}{2}}{\omega} F(0, \omega),$$

(73)

where $\mathscr{P}$ denotes the principal value.

Thus, the real component of the electrical conductivity has the form

$$\mathrm{Re}\,\sigma(\omega) = (2\pi)^3 \frac{\pi \tanh\frac{\beta E}{2}}{\omega} F(0, \omega)$$

(74)

and has the property

$$\mathrm{Re}\,\sigma(\omega) = \mathrm{Re}\,\sigma(-\omega),$$

(75)

which is the consequence of the stationary nature of the random field with respect to time.

It follows from Eq. (74) that measurements of the frequency dependence of the electrical conductivity give direct information on the correlation function; in the region far from the ultra-quantum range (i.e., when $\omega\beta \ll 1$), we simply have

$$\mathrm{Re}\,\sigma(\omega) = 4\pi^4 \beta F(0, \omega).$$

(76)

The imaginary component of the conductivity, which represents the contribution to the permittivity,

$$\mathrm{Im}\,\sigma(\omega) = \frac{ne^2}{m\omega} - \frac{1}{\omega} \int_{-\infty}^{\infty} dE \mathscr{P} \frac{\tanh\frac{\beta E}{2}}{E-\omega} F(0, E)$$

(77)

changes sign under the same conditions when ω is varied.

We shall show how the frequency dependence of the real and imaginary components of the electrical conductivity can be used to deduce the nature of the random field. If the random field is differentiable, the spectral density F(0, ω) can be expanded as a series; then, the transition $\omega \to 0$ gives the same formal equation for the static electrical conductivity as the expression in the absence of random fields:

$$\mathrm{Re}\,\sigma(0) = 4\pi^4 \beta F(0, 0).$$

(78)

At high frequencies, the imaginary component of the electrical conductivity reduces to the contribution of "free" electrons (with renormalized mass)

$$\mathrm{Im}\,\sigma(\omega) = \frac{ne^2}{m\omega^2}$$

only if the spectral density decreases sufficiently rapidly when ω is increased or, in other words, in the absence of "white noise." Otherwise, even in the "nonquantum" region ($\omega\beta \ll 1$)

the permittivity may include a term inversely proportional to temperature. The integral of $\mathrm{Re}\,\sigma(\omega)$ taken over all frequencies is proportional to the variance of the random potential.

We shall consider in detail certain processes, which can be called general relaxation processes, since the usual relaxation is a special case. We shall assume that the dependence of the correlation function (70) on time $t = x_0 - x'$ is described, with an accuracy to within a factor, by a stable characteristic function $f(t)$, so that

$$MF(x, x') = Cf(t). \tag{79}$$

Here, the mathematical expectation symbol represents the averaging over the spatial realization and the constant C is due to such averaging. In the general theory of stationary processes [12], it is shown that the correlation function may be regarded as a characteristic function. We shall be interested not in all the stable characteristic functions, but only in those which satisfy the condition

$$f(t) = f(-t). \tag{80}$$

The condition (80) is naturally associated with the requirement of macroscopic reversibility. The most general form of stable characteristic functions, satisfying the condition (80), is known from [27, 77]:

$$f(t) = e^{-a|t|^\alpha}.$$

Here, $a > 0$ is a parameter; α is a "characteristic exponent," which varies within the limits $0 < \alpha \le 2$. Assuming that $a = |\tau|^{-\alpha}$, we shall write Eq. (80) in the form

$$f(t) = e^{-\left|\frac{t}{\tau}\right|^\alpha}. \tag{81}$$

When $\alpha = 1$, the general relaxation processes reduce to the usual relaxation process for which τ is the relaxation time.

The approximation of the correlation function by a stable characteristic function can be understood in various ways. First of all, stable distributions appear, as already demonstrated, when we sum an infinitely large number of independent and identically distributed terms. In our case, it means that the relaxation is, in fact, the result of a large number of small and independent effects or it may be represented as the result of such effects. However, we must bear in mind the definitions of stable distributions in terms of characteristic functions: for any values of a_1 and a_2, where $a_1 > 0$, $a_2 > 0$, there should exist such values of $a > 0$ and b, that

$$f(a_1 t) f(a_2 t) = f(at) e^{ibt}. \tag{82}$$

We are interested in the case b = 0. The definition (82) means that two independent (i.e., sufficiently far apart in space or in time) relaxation processes, in each of which the time scale can be altered arbitrarily, may be considered as one process with a suitably selected time scale.

The importance of the general relaxation processes in the transport phenomena in solid solutions is due to the fact that the usually employed power dependence of the relaxation time or mobility is equivalent to the assumption of stability. In order to show this, we shall start from Kubo's formula for the static electrical conductivity [71]:

$$\sigma = A \int\limits_0^\infty dt \int\limits_0^\beta f(t + ih\lambda)\, d\lambda, \tag{83}$$

where $f(t)$ is, by definition, given by Eq. (81), and the constant A is due to the averaging over the spatial variables. Introducing polar coordinates $\rho^2 = t^2 + h^2\lambda^2$, $\theta = \tan^{-1}(t/h\lambda)$, we can transform Eq. (83) into

$$\sigma = A \frac{\tau^2}{\alpha} \int\limits_0^{\pi/2} d\Theta\, \gamma\left(\frac{2}{\alpha}, \left(\frac{\beta h}{\tau \sin\theta}\right)^\alpha\right), \tag{84}$$

$$\gamma(a, x) = \int\limits_0^x e^{-t} t^{a-1}\, dt,$$

where $\gamma(a, x)$ is an incomplete γ function.

Integration by parts gives

$$\sigma = A \frac{\tau^2}{\alpha} \left\{ \frac{\pi}{2} \gamma\left(\frac{2}{\alpha}, \left(\frac{\beta h}{\tau}\right)^\alpha\right) + \alpha \left(\frac{\beta h}{\tau}\right)^2 \int\limits_0^{\pi/2} \frac{\theta\, d\theta}{\tan\theta} e^{-\left(\frac{\beta h}{\tau \sin\theta}\right)^\alpha} \right\}. \tag{85}$$

Outside the ultra-quantum range, we shall assume that

$$\frac{\beta h}{\tau} \ll 1. \tag{86}$$

For not-too-small values of α this requirement implies $(\beta h/\tau)^\alpha \ll 1$ and it allows us to omit the second term in Eq. (85), which is small compared with the first. An expansion of the incomplete γ function in powers of x

$$\gamma(a, x) = \sum\limits_{n=0}^\infty \frac{(-1)^n}{n!} \frac{x^{a+n}}{a+n}$$

gives

$$\sigma = A \frac{\tau^2}{2} \left(\frac{\beta h}{\tau}\right)^\alpha. \tag{87}$$

In the usual expression for the electrical conductivity

$$\sigma = \frac{ne^2}{m} \tau_0$$

the average relaxation time τ_0 depends on temperature as $\beta^{3/2}$ in the case of scattering by the acoustical vibrations, and as $\beta^{1/2}$ in the case of scattering by the optical vibrations, but is independent of temperature for the scattering by neutral impurity atoms. It follows from Eq. (87) that these three cases are covered by our model of the general relaxation processes, and that the characteristic exponent α determines the temperature dependence of the mobility. This result makes it possible to consider any observed temperature dependence of the mobility of the β^α type as being due to some general relaxation process with a suitable characteristic exponent. The limitation $0 < \alpha \leq 2$ imposed on the characteristic exponent means that the model with a stable characteristic function is not universal. Thus, the scattering by ionized impurities is not described by this model, which means that, in agreement with the Coulomb nature of scattering on ionized impurities, we cannot in this case regard the relaxation process as being due to a large number of small effects.

As already mentioned, the differentiability of the random field can be deduced from the existence of the second derivative of the correlation function at t = 0. In our case, the second derivative of the expression

$$f(t) = e^{-\left|\frac{t}{\tau}\right|^{\alpha}}$$

exists at t = 0 only in two cases: when $\alpha = 1$ and $\alpha = 2$. For fractional values of α, the random fields are not differentiable. This means that during each infinitely short (in the physical sense) time interval an electron suffers an infinite number of effects. Such effects are the cause of the diffuse scattering.

For the sake of completeness, we must mention also that stable distributions can be used to consider momentum distributions.

PART II

EXPERIMENTAL DATA

7. ATOMIC SOLID SOLUTIONS

Ge — Si solid solutions. The phase diagram of the Ge—Si system has been investigated theoretically by Romanenko and Ivanov-Omskii [82]. They have shown that to make the theoretical liquidus and solidus curves agree with the observations it is necessary to assume that the interaction between like atoms (Ge—Ge and Si—Si) should be weaker than that between unlike atoms (Ge—Si), and, therefore, solid solutions should have a tendency to ordering. The same conclusion has been reached by Braunstein et al. [106] and by A. M. Toksen on the basis of thermal conductivity measurements at low temperatures. This is obviously a manifestation of the relationship referred to in Section 1: a solid solution is "ideal," i.e., it does not decompose into two phases in a finite time interval, if the structure of the electronic spectra of the components of the solution are similar. Otherwise, a solid solution is, in fact, in a metastable state. It is very likely that this relationship applies also in the opposite sense: a tendency to form solid solutions is exhibited primarily by the compounds which have similar, at least qualitatively, band structures.

An interesting property of Ge—Si solid solutions is the kink in the concentration dependence of the optically measured forbidden bandwidth (Fig. 4), observed by Johnson and Christian [157]. An interpretation of this kink, given by Herman et al. [154, 155], is as follows. It is assumed (Fig. 5) that the hole bands of Ge—Si solid solutions remain, at all concentrations, similar to the hole bands of the pure elements, while the electron bands are deformed smoothly by the variation of the concentration. If we neglect the spin—orbital splitting, we find that electrons in silicon have two types of energy minimum: at the point L_1 and along the <100> direction, and the former lies higher than the latter. However, in germanium, the minimum at the point L_1 lies below the minimum along the <100> direction. The positions of both minima coincide at a concentration of about 15 at.% Si. Thus, the kink is simply due to the fact that at low silicon concentrations electrons are transferred to minima of the <111> type, as in pure germanium, while at high silicon concentrations they are transferred to the <100> minima, as in pure silicon.

These conclusions agree with the magnetoresistance measurements [138]. The results of such measurements at silicon concentrations of less than 10 at.% may be interpreted on the assumption that there are electron ellipsoids, oriented along the <111> direction, whereas, at silicon concentrations of 10-17 at.% a different anisotropy is apparent, which is not in agreement with the assumption of ellipsoids of one type only. The introduction of additional ellipsoids, oriented along the <100> direction makes the measurements agree with the theory; the energy gap between ellipsoids of different types is found to depend on the concentration. At silicon concentrations higher than 23 at.%, the results can be accounted for solely by ellipsoids oriented along the <100> direction with effective masses close to the masses observed in pure silicon. The same conclusions follow from the measurements of the magnetic susceptibility

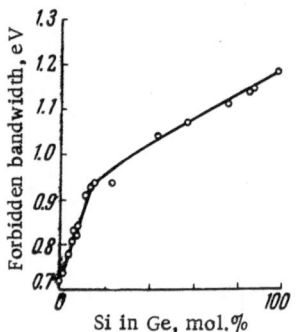

Fig. 4. Dependence of the forbidden bandwidth of Ge−Si solid solutions on the composition.

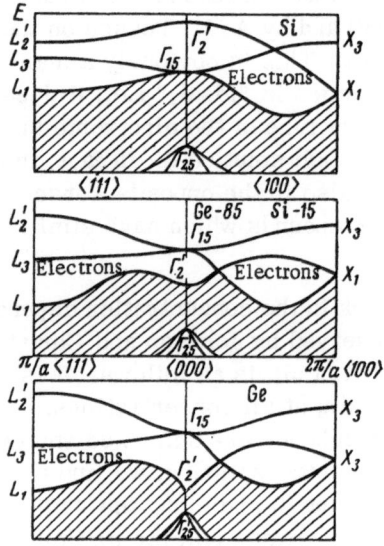

Fig. 5. Structure of the bands (without the spin−orbital splitting) along the <100> and <111> directions for silicon, for 15% solution of silicon in germanium, and for germanium.

[39]. Levitas [171] investigated the mobility of holes in Ge−Si solutions at 77-300°K and found that, if the temperature dependence of the mobility was described parametrically by $A_1 T^{-\alpha}$, then both A_1 and α had a minimum at 30-40% of silicon. Glicksman [140] investigated in detail the 0-30 at.% Si range. After subtracting from the observed mobility the component due to the scattering by ionized impurities, Glicksman found that the scattering occurred mainly on phonons and that at concentrations lower than 10 at.% Si, electrons of the <111> type were scattered (as in pure germanium), whereas, at concentrations higher than 20 at.% Si, electrons of the <100> type were scattered. In the intermediate range of concentrations, the two types of scattering could be combined to agree with the experimental data. However, that contribution to the mobility which was inversely proportional to the product of the concentrations was found to depend on temperature as $T^{-0.8}$. In the interpretation of these results, it would seem reasonable to use the representation of the general relaxation processes. Changes in the optical and electrical properties under pressure were also investigated [187]. The optical width of the forbidden band E_g was measured up to 7000 kg/cm². The derivative $(\partial E_g / \partial P)_T$, plotted against the concentration, had a kink at 10-20 at.% Si, which was in qualitative agreement with the band scheme proposed by Herman. However, some details, in particular the form of the absorption band edge, did not agree fully with this scheme.

Measurements of the optical reflection of solid solutions, as a function of the wavelength and concentration, provide another method of investigating the band structure. It is known that germanium and $A^{III}B^V$ compounds have a reflection peak lying approximately in the 2 to 2.3 eV range, whereas silicon has a similar peak at 3.3 eV. This peak is ascribed to direct transitions between the points L (points on the boundary of the Brillouin zone along the <100> direction) and the doublet structure of the peak is ascribed to the spin−orbital splitting. Similar features are also observed for Ge−Si solutions. Measurements of the transition energy E_L as a function of the composition have shown [208] that E_L of solutions with high concentrations of germanium is in qualitative agreement with the band structure just referred to, but the kink in the dependence of the transition energy E_L on the concentration is observed at 79 at.% Si. This suggests that at high silicon concentrations the point L_3 lies lower than the point L_1.

Reports have also been published [155] of experimental investigations of deep traps formed by copper and gold in solid solutions in the same way as they are formed in pure germanium. The position of an acceptor level of gold at a depth of 0.15 eV has been found to vary nonlinearly with the concentration; a donor level at 0.05 eV shifts linearly away from the valence band when the concentration of silicon is increased from 0 to 12 at.%. All these observations cannot be explained simply by the variation of the permittivity, but it is necessary to allow more rigorously for the interaction between traps and their environment.

Se — Te solid solutions are formed at all concentrations [182].

Ge — Sn solid solutions have been investigated to find whether elements of the tin group greatly affect the electrical properties. It has been found that tin atoms are neutral at concentrations of 10^{20} atoms/cm^3. Measurements of the carrier lifetime have also shown that at these concentrations they do not act as recombination centers [211].

8. SOLUTIONS OF $A^{III}B^{V}$ COMPOUNDS

The band structures of the $A^{III}B^{V}$ compounds are basically similar, as indicated, in particular, by the similarity of the dependence of the absorption coefficients on the wavelength, shown in Fig. 6. However, the absolute positions of the various maxima and critical points observed experimentally can differ very widely, as indicated by the kinks in the concentration dependences of the bandwidths (compare with Ge−Si solid solutions) or by a direct comparison of the band structures, deduced from all the available experimental data. Figure 7 shows the band structure of GaAs, deduced from the data on optical transitions, effective masses, theoretical calculations of the spin−orbital splitting, etc. Arrows indicate those optical transitions which can be seen in Fig. 6. Figure 8 shows the band structure of GaP. Comparison of Figs. 7 and 8 shows that only the general topological characteristics of the bands of these two compounds are the same (the total number of critical points is probably the same), but the quantitative properties, for example, the effective masses or even the positions of the critical points, show that the band structures of these two compounds differ very considerably.

GaAs − GaP solid solutions are formed at all concentrations. The dependence of the absorption band edge on the concentration (Fig. 9) has a kink approximately at 50 at.%; this kink is due to the fact that the forbidden bandwidth of GaAs is governed by a direct transition, whereas that of GaP is governed by an indirect transition to the point X (this point lies on the boundary of the Brillouin zone along the <100> direction). A more detailed investigation of direct and indirect transitions has been carried out by Spitzer and Mead [203]; their results are presented in Fig. 10. An even more detailed investigation of the optical absorption of these solid solutions has been carried out by Abagyan, Lishina, and Subashiev [1]. The straight lines 1, 2, and 3 in Fig. 11 represent the concentration dependences of the transitions at 240°K to the points <000>, <111>, and <100> in the conduction band. The dashed line in Fig. 11 is drawn through the points corresponding to a rapid increase in the absorption coefficient. Abagyan et al. have concluded that the mobility of electrons in GaAs−GaP solutions should, other conditions being equal, depend strongly on the composition, remaining high (as for GaAs) in the region where electrons are located mainly at the <000> point, and falling rapidly in the transition to compositions for which the bottom of the conduction band is determined by the <100> states. The properties of the investigated solutions have been used by Abagyan et al. to refine the band structure of the components.

InAs − InP solid solutions. Measurements of the optical forbidden bandwidth E_g and the electron mobility as a function of the composition in the 0-60 mol.% InP range, have shown that E_g varies linearly with the concentration of InP. This is interpreted on the basis of the similarity of the band structures of InAs and InP not only with respect to the number of critical points, but also with respect to their positions. The mobility decreases rapidly when InP is added, reaching 11,000 $cm^2 \cdot V^{-1} \cdot sec^{-1}$ at 20 mol.% InP; but at higher concentra-

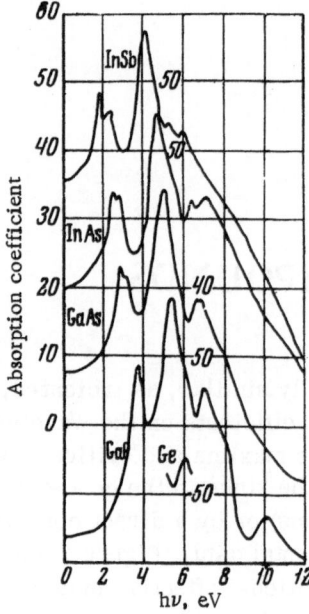

Fig. 6. Absorption coefficients of some
III–V compounds.

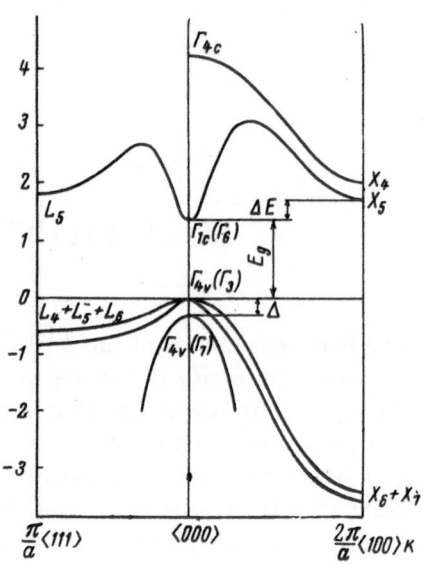

Fig. 7. Band structure of GaAs according
to [168].

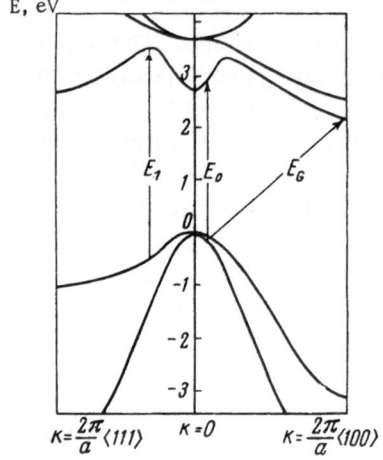

Fig. 8. Band structure of GaP according
to [237a].

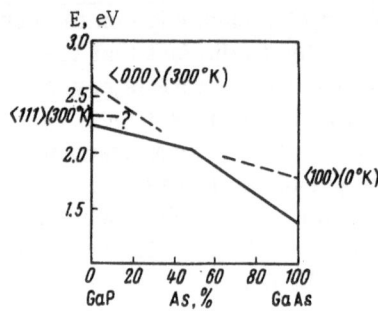

Fig. 9. Absorption band edge of GaP–GaAs
solid solutions as a function of composition,
according to [168].

tions of the phosphide, it decreases more slowly. The ratio of the electron and hole mobilities is very high (about 100) at all concentrations. The effective masses increase from InAs to InP. Ehrenreich [130] has calculated the mobilities of their solid solutions from the mobilities of the pure components on the assumption that both compounds have the same band structure. The results, which are in reasonable agreement with the experimental data, indicate that the scattering by inhomogeneities is weak and the ionic component of the chemical bonds important. Measurements of the thermal conductivity [104, 215] have been carried out at room temperature up to 60 mol.% InP, and from room temperature to 800°C at 0–40 mol.% InP. As in all cases of

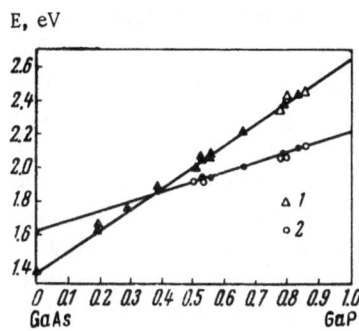

Fig. 10. Absorption band edge of GaAs−GaP solid solutions, determined from the photo emf [203]: 1) direct transitions; 2) indirect transitions.

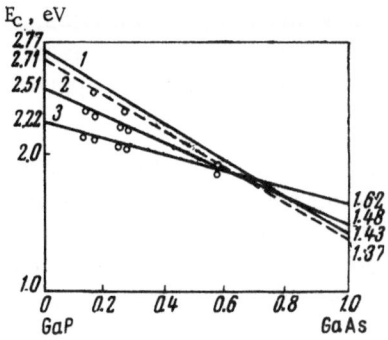

Fig. 11. Transition energy as a function of concentration for GaP−GaAs solid solutions [1].

this kind, the minimum thermal conductivity occurs at an approximately equimolar composition. The results given in these two papers [104, 215] differ somewhat in the intermediate range of concentrations, but they give the same values of the thermal conductivity for the pure compounds. The temperature dependence of the thermal conductivity of the pure compounds is found to be much stronger than for the solid solutions. The thermoelectric figure of merit Z has been determined as a function of temperature in the 0-40 mol.% InP range; its maximum has been found to lie at 650-750°C and 10 mol.% InP. The magnetic susceptibility and the Hall coefficient have also been measured [109].

GaAs − InAs solid solutions. The thermal conductivity, the Hall electron mobility, the optical widths of the forbidden band, and the thermoelectric power of these solid solutions have been investigated [105]; the investigated samples were not uniform for all the compositions. The dependence of the optical forbidden bandwidth on the concentration is concave in the direction of the concentration axis. For more uniform samples [88], this dependence approaches linearity. The thermal conductivity at room temperature decreases rapidly when the concentration of either component is increased, reaching a minimum at the equimolar composition [88].

AlSb − GaSb solid solutions. The optical forbidden bandwidth of these solutions varies almost linearly with the concentration. Burdiyan and Kolomiets [23] have measured the electrical conductivity and the Hall effect at 100-900°K; at room temperature, the mobility has been found to be 75-250 cm$^2 \cdot$ V$^{-1} \cdot$ sec^{-1} (the mobility in GaSb is 1400 cm$^2 \cdot$ V$^{-1} \cdot$ sec^{-1} and in AlSb, 420 cm$^2 \cdot$ V$^{-1} \cdot$ sec^{-1}). Miller et al. [177] have found that the mobility in samples with 40 mol.% GaSb above 700°K depends on the temperature as T$^{-3/2}$, which indicates that phonon scattering is dominant at these temperatures. The forbidden bandwidth, found from the temperature dependence of the electrical conductivity, is close to the optically determined value. Investigations of the thermoelectric properties at 120-900°K have shown that the thermoelectric power has a maximum at 350-500°K. The effective masses are maximal at 40-50 mol.% GaSb.

AlSb − InSb solid solutions. The resistivity and the Hall coefficient of these solid solutions have been measured as a function of temperature for samples containing 10, 25, and 50 mol.% AlSb [4]. At room temperature, the hole mobility decreases from 600 cm$^2 \cdot$ V$^{-1} \cdot$ sec^{-1} for 10 mol.% AlSb to 80 cm$^2 \cdot$ V$^{-1} \cdot$ sec^{-1} for 50 mol.% AlSb. For the same compositions, the parameter α in the temperature dependence of the mobility T$^{-\alpha}$ is 1.67 and 1.1, respectively, indicating that the scattering by inhomogeneities predominates. Observations on the Nernst−Ettingshausen effect [3] support this conclusion, but they also indicate that when the temperature is increased, the lattice scattering becomes more and more important.

GaSb − InSb solid solutions. These solutions demonstrate clearly the limitations of the virtual crystal model: the optical value of the forbidden bandwidth, deduced from numerous and independent measurements at room temperature, is a nonlinear function of the concentration and has a kink approximately at 75 mol.% GaSb, in spite of the fact that the band structures of the components GaSb and InSb are similar. The temperature coefficient of the optical forbidden bandwidth varies smoothly over the full range of concentrations and, consequently, the optical bandwidth at absolute zero, found by extrapolation, also has no kinks and may even vary linearly with the concentration. The electron mobility decreases rapidly when GaSb is added to InSb: it falls to a value of 10,000 $cm^2 \cdot V^{-1} \cdot sec^{-1}$ and remains almost constant up to 85 mol.% GaSb; at still higher concentrations of GaSb, it falls to a value of 4500 $cm^2 \cdot V^{-1} \cdot sec^{-1}$, which is typical of GaSb, so that over the whole range of concentrations the mobility of the solutions is higher than the mobility of the component GaSb. The ratio of the mobilities behaves similarly. For GaSb concentrations higher than 60 mol.%, the electrical measurements give two different bandwidths: E_1 at low temperatures, and E_2 at high temperatures, with a transition between these widths at 650-800°K. For compositions with 60-85 mol.% GaSb, the optical bandwidth, extrapolated to absolute zero, is close to E_2, but near GaSb it is close to E_1. Evidently, in this case the change in the composition alters the band structure so that at 75 mol.% GaSb the bandwidth is determined not by the <000> minimum, as in pure components, but by some other minimum. Right up to about 600°K, samples having compositions of 20-60 mol.% GaSb have an electron mobility which is proportional to $T^{-0.8}$ (as for Ge—Si solid solutions). This indicates that the scattering by inhomogeneities predominates, although the power exponent of the temperature also includes a phonon contribution. Measurements of the thermal conductivity have shown that, as usual, there is a minimum at an approximately equimolar composition.

Among solid solutions of $A^{III}B^V$ and other compounds, we must mention the InAs—CdTe system, which has been investigated up to 53.5 mol.% CdTe and up to 38 mol.% InAs; in the intermediate range, the solutions split up into two phases. The Hall coefficient and the thermal conductivity of these solutions have been measured. The lattice component of the thermal conductivity has been found to decrease when the composition approaches the equimolar ratio.

$InSb-In_2Te_3$ solutions have been investigated by Aliev and Dzhangirov [5].

9. OTHER SOLID SOLUTIONS

ZnS − CdS and ZnS − ZnSe solid solutions are interesting primarily because of their luminescence properties. The absorption band edge of ZnS−CdS solutions varies linearly with the composition [138].

ZnS − HgS solid solutions [169] have a forbidden bandwidth, measured up to 57 mol.% HgS, which varies linearly from 3.3 to 1.4 eV with the composition; further increase in the concentration of HgS should produce a kink, because extrapolation gives a negative value of the forbidden bandwidth.

CdTe − HgTe solid solutions [170]. The Hall coefficient and the electrical conductivity of these solid solutions have been measured as a function of temperature. High values of the Hall mobility (about 23,000 $cm^2 \cdot V^{-1} \cdot sec^{-1}$) have been obtained for samples with 90% HgTe. The mobility depends on temperature as T^{-2}. The optical absorption and the photoconductivity of these solutions have also been measured. The thermal conductivity has a minimum near the equimolar composition, and this minimum has been observed to vary weakly with temperature right up to 250°C.

CdTe − ZnTe solid solutions [67]. Measurements of the optical absorption indicate a linear variation of the forbidden bandwidth with the composition from 1.45 eV for CdTe to 2.1 eV for ZnTe, indicating that the band structure of the two components is the same. Measurements of the thermal conductivity [59] (mainly of the lattice component) reveal a broad minimum near the equimolar composition.

CdSe − ZnSe solid solutions [99]. These solid solutions are not formed over the whole range of compositions: there is a narrow interval (31–33.5 mol.% CdSe), in which the solid solutions decompose into two phases. The forbidden bandwidth values, found from the optical absorption and reflection and from the internal photoeffect, are found to agree well; The forbidden band varies from 2.65 eV for ZnSe to 1.7 eV for CdSe. Measurements of the temperature dependence of the electrical conductivity give completely different results: according to the conductivity data, the forbidden bandwidth varies in different ways on both sides of the two-phase interval. This behavior of the electrical conductivity is attributed to the different scattering mechanisms.

HgSe − HgTe solid solutions [79] have been investigated by Nikol'skaya and Regel'. These solutions exhibit an interesting feature: a mobility peak is observed near the equimolar composition. Later measurements have not confirmed this result [20]. According to [20], the increase in the electrical conductivity near the equimolar composition is due to an increase in the number of carriers, while the mobility has a minimum in this region. This minimum is not very deep (16,000 $cm^2 \cdot V^{-1} \cdot sec^{-1}$, compared with 20,000 $cm^2 \cdot V^{-1} \cdot sec^{-1}$

for HgSe and 23,000 $cm^2 \cdot V^{-1} \cdot sec^{-1}$ for HgTe). The same paper reports a determination of the effective electron mass, which has a maximum at the equimolecular composition and is approximately one and one-half times larger than the effective masses of the components (0.030m for HgSe and 0.033m for HgTe). Measurements of the thermoelectric power and of the Hall effect [193] yield somewhat different masses for the solutions and their components (0.044m and 0.030m, respectively); it has been concluded from these measurements that the conduction band is not parabolic, and that the dominant scattering mechanism at room temperature is the scattering by phonons. More detailed investigations near the equimolar region have shown that these results cannot be interpreted within the framework of a simple two-band model, and that it is necessary to assume the existence of one conduction band and two valence bands [49]. It has been concluded, moreover, that the properties are more semimetallic than semiconducting.

SnS — SnSe solid solutions. These solutions are interesting because their thermal conductivity at the equimolar composition falls to only 70% of the value of the thermal conductivity for the pure components.

Other solid solutions have been investigated either as part of a physicochemical analysis or in connection with applications. Even a qualitative interpretation of the results obtained is difficult because of the complexity of the investigated materials and the absence of a complete theory of the behavior of quasi-particles in random fields.

10. GLASSY SEMICONDUCTORS

We have already mentioned that the methods presented in Part I are applicable not only to solid solutions. Glassy semiconductors are among substances whose electronic properties can also be treated within the framework of the theory of quasi-particles in spatially inhomogeneous random fields. In fact, in the effective mass approximation the long-range order is allowed for phenomenologically through the effective mass; and if we are interested in the minima in a band, the aspects of a problem typical of a crystal are ignored. Equally well, we can consider the behavior of carriers in a glass or in a polymer, regarding carriers with a quadratic dispersion law as "primary," and then allowing for the presence of the random field, using one of the variants of the perturbation theory. For this reason, it seems convenient to give here a brief review of the properties of glassy semiconductors, particularly because solid solutions and glasses are one-phase systems of variable composition, which have similar macroscopic properties. For example, in both cases, the electrical conductivity depends exponentially on temperature in the usual manner. However, there are also differences: for example, no impurity levels or bands have yet been observed in glasses, and trapping centers are only sometimes observed at low temperatures. It is quite likely that the absence of impurity states is due to the simultaneous effects of two factors: the pushing out of "impurity levels" into a band when the carrier density is increased, and the presence of a "tail" in the density of states.

In spite of the qualitative similarity of the electronic properties of glasses and ideal crystals, quantitative parameters of even the same substance may differ considerably in the crystalline and glassy states. Thus, according to Kolomiets et al. [60], the conductivity of glasses of the $As_2Se_3-As_2Te_3$ system increases by a factor of 10^{10} on crystallization. The mobility of carriers in glasses is probably very low, so that once again we meet complications typical of low-mobility semiconductors.

A. F. Ioffe has advanced a hypothesis, based on numerous experimental observations, that the electronic properties of condensed media (including crystals and glasses) are governed by their short-range order. It has been shown subsequently [6] that this hypothesis is based essentially on a mathematical theorem which says that, under very general conditions, the eigenvalues of an operator (for Schrödinger's equation or for the effective wave equation) vary slightly when the operator or the boundary conditions are altered a little. This theorem is based on a consideration of cells constructed, in the same way as in the Wigner–Seitz method, by bisecting the lines joining an atom with its nearest neighbors. If the short-range order in a crystal and in an amorphous substance is the same, the shapes of the cells are also the same. However, the solution of a differential equation can be found piecewise by joining a number of particular solutions and their derivatives so as to ensure continuity. Thus, a cell in a crystal and a cell in an amorphous substance are geometrically similar and the difference between the

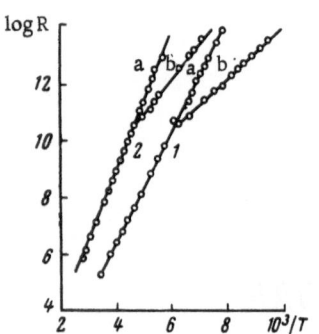

Fig. 12. Logarithm of the resistance as a function of the reciprocal of temperature for glassy As_2Te_3 and As_2SeTe_2.

statements of the problem in the two cases consists only of a slight change of the potential and of the boundary conditions, due to the influence of neighbors outside the limits of the first coordination sphere. If this influence can be regarded as a perturbation, the spectrum of an amorphous substance with a given short-range order should, in the usual perturbation theory meaning, differ little from the spectrum of a crystal with the same short-range order. However, it is very important to note that the absence of the long-range order destroys completely the concept of a zone as an assembly of states which are transformed in accordance with the translation group representations.

Similar conclusions follow from a consideration of a problem of the spectrum in the tight-binding approximation for any number of coordination spheres, which can be as large as we please (using either the LCAO or the LCMO method) [8]. The secular equations obtained in these methods may have roots, concentrated in a certain finite range of energies, both in the presence and in the absence of spatial periodicity. Therefore, the translational invariance is only a necessary but not a sufficient condition for such a concentration of energies when one band is considered. If we bear in mind that the band overlap is not associated with the translational invariance (both metals and insulators are translationally invariant in an ideal crystal approximation), we must realize that the translational invariance is not a sufficient condition for the existence of a band. In other words, the translational invariance is in no way related to the vanishing of the density of states.

Interesting experimental data on the optical properties of glassy semiconductors have been reported by Zorina (cf. [52], which also gives further references). Interpreting the observed absorption in glassy As_2S_3 and As_2S_5, as in the case of ideal crystals, Zorina came to the conclusion that direct and indirect transitions are possible in glasses and that, consequently, a complex band structure with several minima is not a specific characteristic of the ideal crystal state. The final solution of the problem of the correctness of such an interpretation and of the conclusions that follow from it may be obtained after the development of a general theory of electronic properties of glasses. Such a theory should allow for the low values of the mobility (lifetimes) and, consequently, for the strong interaction of carriers with random fields. This strong interaction makes inapplicable those variants of the theory in which the interaction is regarded as a perturbation.

Experimental investigations of the transport properties of glasses (lifetimes, etc.) are difficult because of their high resistivities. One of the few exceptions is the $Te_2SeAs_2(Se, Te)_3$ system, investigated by Ivkin, Kolomiets, and Lebedev [60]. They have found that at a room temperature the mobility increases by a factor of 4 when the Te concentration is increased from 15 to 37.5%, and that this mobility is $(2-8) \cdot 10^{-2}$ $cm^2 \cdot V^{-1} \cdot sec^{-1}$. The temperature dependence of the Hall coefficient has also been measured for one of the compositions of the system. The Hall mobility at temperatures between −30 and +70°C has been found to be independent of temperature. Observations of the carrier drift have shown that these materials contain hole and electron capture centers, and that the effectiveness of these centers is approximately the same for both types of carrier.

Interesting measurements of the temperature dependence of the electrical conductivity of glasses have been carried out by Gubanov and Mazets [43]. Figure 12 shows the dependences of the logarithm of the resistance on the reciprocal of temperature for glassy As_2Te_3 (1) and As_2SeTe_2 (2) in darkness (a) and after illumination for 5 min at $-196°C$ (b). The kinks in the temperature dependences of the electrical conductivity, due to the illumination, provide a direct proof of the presence of "impurity" states in glasses at low temperatures.

CONCLUSIONS

It follows from our review of the experimental data on semiconducting solid solutions that direct analysis of these data within the framework of the theory presented in Part I is difficult for at least two reasons. First, experiments carried out so far have not been intended to determine the probability characteristics of the random potential in solid solutions. Therefore, in these experiments, solid solutions have been considered simply as ideal crystals. Secondly, such ideal crystals have been investigated using only the one-electron approach and, therefore, experimenters have not been interested in many important effects outside this approach, in particular those due to the screening of the random potential. Again, a characteristic property of solid solutions has been ignored. Since these special properties of solid solutions will sooner or later become important in detailed studies, we must make some general remarks on the behavior of carriers in random fields.

In the effective mass approximation, carriers in an ideal crystal are described by a quadratic dispersion law

$$\varepsilon = p^2/2m$$

or by several expressions of this type. This expression is formally identical with the kinetic energy of a free particle. Therefore, any potential external to a carrier may be regarded as equal to zero. In a solid solution, the situation is different: local deviations from the number of atoms of a given kind, producing a potential at a given point, force us to consider the potential as a random function. Moreover, we must remember that carriers in a solid have quantum-mechanical properties and are described by wave functions. Any form of random potential may give rise to local states at those points where there are sufficiently deep potential wells, which are due, for example, to a local accumulation of positive charge. In this case, a carrier is captured by a well, forms a local state, and either completely or partly neutralizes the positive charge. Small accumulations of positive charge do not give rise to local states (this case is typical of solid solutions). However, in the vicinity of a well, a wave function of an electron still forms an "antinode" (the probability of finding an electron near a positive charge increases), which equalizes the potential. Obviously, the larger the number of carriers, i.e., the higher the temperature, the greater the degree of leveling of the potential. We thus find that the energy spectrum depends on temperature.

These considerations are based on the assumption that the concept of carriers with a "primary" dispersion law is justified, or, which is equivalent, they are based on the assumption that the random field is weak and can be regarded as a perturbation. In view of the increase in the screening effect when the temperature is increased, this assumption can be regarded as justified at sufficiently high temperatures.

The influence of a random field on the observed quantities is manifested not only by changes in the "intrinsic" properties of carriers, such as the effective mass, but also indirectly. Thus, on going over from ideal crystals to solid solutions, we find that the lines of the optical transitions become broader, the edge of the absorption band becomes wider, etc. The presence of a random field is felt particularly strongly in the electrical conductivity, and in transport processes in general. Its influence can be regarded simply as the "friction" of a carrier against the random potential "roughness" with a consequent reduction in the mobility; in view of the quantum nature of carriers, this model is of very limited use.

In experimental investigations of the influence of the random field on electronic properties, it would be desirable to determine the frequency dependence of the electrical conductivity and to measure transport phenomena at low temperatures. Further experimental and theoretical considerations along these directions may give more detailed information on random fields in solid solutions.

REFERENCES

1. S. A. Abagyan, A. V. Lishina, and V. I. Subashiev, Conduction band minima of the GaAs−GaP system, Fiz. Tverd. Tela, 6:2852 (1964).

2. S. A. Abagyan, V. I. Subashiev, and S. P. Singkhal, Nature of ultraviolet reflection maxima of GaP and $GaAs_xP_{1-x}$, Fiz. Tverd. Tela, 6:3186 (1964).

3. Ya. Agaev, O. V. Emel'yanenko, and D. N. Nasledov, Investigation of the Nernst−Ettingshausen thermomagnetic effects in solid solutions of the InSb−AlSb system, Fiz. Tverd. Tela, 3:194 (1961).

4. Ya. Agaev and D. N. Nasledov, Some electrical properties of the AlSb−InSb system, Fiz. Tverd. Tela, 2:826 (1960).

5. M. I. Aliev and A. Yu. Dzhangirov, Investigation of thermal and electrical properties of $InSb−In_2Te_3$ alloys, Fiz. Tverd. Tela, 8:2415 (1964).

6. A. B. Almazov, Influence of short-range order on electrical properties of matter, Fiz. Tverd. Tela, Sbornik (collection) 2:158 (1959).

7. A. B. Almazov, Expansion of dispersion laws of quasi-particles in crystals in terms of theta functions, Fiz. Tverd. Tela, 4:1293 (1962).

8. A. B. Almazov, Spectrum of electrons in a polymer chain, Doklady Akad. Nauk SSSR, 138:863 (1961).

9. A. B. Almazov, Theory of quasi-classical fermions in random fields and its application to electrons in glasses and inorganic polymers in bulk, Fiz. Tverd. Tela, 5:1320 (1963).

10. A. I. Ansel'm, Introduction to the Physics of Semiconductors. Izd. Akad. Nauk SSSR, Moscow (1962).

11. B. V. Baranov and N. A. Goryunova, Preparation of homogeneous solid solutions of the AlSb−InSb system, Fiz. Tverd. Tela, 2:284 (1960).

12. M. S. Bartlett, An Introduction to Stochastic Processes, 1955 [Russian translation]. IL, Moscow (1958); A. A. Sveshnikov, Application Methods in the Theory of Random Functions, Sudpromgiz (1961).

13. M. L. Belle, Optical reflection of PbSe, PbTe, and of their solid solutions, Fiz. Tverd. Tela, 7:606 (1965).

14. O. V. Bogorodskii, A. Ya. Nashel'skii, and V. Z. Ostrovskaya, X-ray diffraction investigation of InAs−InP solid solutions, Kristallografiya, 6:119 (1961).

15. V. L. Bonch-Bruevich and S. V. Tyablikov, Green's Function Method in Statistical Mechanics, Fizmatgiz, Moscow (1961).

16. V. L. Bonch-Bruevich and A. G. Mironov, Influence of impurities on the energy spectrum of carriers, Fiz. Tverd. Tela, 3:3009 (1961).

17. V. L. Bonch-Bruevich, Theory of heavily doped semiconductors, Fiz. Tverd. Tela, 4:2660 (1962).

17a. V. L. Bonch-Bruevich, Problems in the Theory of Heavily Doped Semiconductors (collection: Itogi Nauki), Izd. VINITI (1965).

18.   A. S. Borshchevskii, I. I. Burdiyan, É. Yu. Lubenskaya, and E. V. Sokolov, Phase diagram of AlSb−GaSb, Zh. Neorg. Khim., 4 : 2824 (1959).

19.   A. S. Borshchevskii and D. N. Tret'yakov, Use of vibrational mixing in synthesis of semiconducting materials, Fiz. Tverd. Tela, 1 : 1483 (1959).

20.   B. Ya. Brach, V. V. Zhdanova, and E. Ya. Lev, Thermoelectric properties of the HgSe−HgTe system, Fiz. Tverd. Tela, 3 : 785 (1961).

21.   I. I. Burdiyan, Some additional information on AlSb−GaSb solid solutions, Fiz. Tverd. Tela, 1 : 1360 (1959).

22.   I. I. Burdiyan and A. S. Borshchevskii, Preparation and some properties of solid solutions of the AlSb−GaSb system, Zh. Tekh. Fiz., 28 : 2684 (1958).

23.   I. I. Burdiyan and B. T. Kolomiets, Investigation of the conductivity and of the Hall effect of solid solutions of the AlSb−GaSb system, Fiz. Tverd. Tela, 1 : 1665 (1959).

24.   I. I. Burdiyan, N. A. Rozneritsa, and G. I. Stepanov, Thermoelectric properties of solid solutions of the AlSb−GaSb system, Fiz. Tverd. Tela, 3 : 1879 (1961).

25.   N. I. Vitrikhovskii and I. B. Mizetskaya, Preparation of mixed CdS−CdTe single crystals from the vapor phase and some of their properties, Fiz. Tverd. Tela, 1 : 397 (1959).

26.   N. I. Vitrikhovskii and I. B. Mizetskaya, Mixed ZnS−CdS single crystals and some of their properties, Fiz. Tverd. Tela, 2 : 2579 (1960).

27.   B. V. Gnedenko and A. N. Kolmogorov, Limit Theorems for Sums of Independent Random Quantities, Moscow (1949).

28.   G. N. Gordyakova, G. V. Kokosh, and S. S. Sinani, Investigation of thermoelectric properties of Bi_2Te_3−Bi_2Se_3 solid solutions, Zh. Tekh. Fiz., 28 : 3 (1958).

29.   N. A. Goryunova, Substitutional Solid Solutions in Compounds with the Zinc Blende Structure, Izd. Akad. Nauk SSSR (1955).

30.   N. A. Goryunova and B. V. Baranov, Solid solutions in the AlSb−InSb system, Doklady Akad. Nauk SSSR, 129 : 839 (1959).

31.   N. A. Goryunova and I. I. Burdiyan, Solid solutions in the AlSb−GaSb system, Doklady Akad. Nauk SSSR, 120 : 1031 (1958).

32.   N. A. Goryunova and V. S. Grigor'eva, Gallium arsenides, Zh. Tekh. Fiz., 26 : 2157 (1956).

33.   N. A. Goryunova, V. S. Grigor'eva, B. M. Konovalenko, and S. M. Ryvkin, Photoelectric properties of some compounds with the zinc blende structure, Zh. Tekh. Fiz., 25 :1675 (1955).

34.   N. A. Goryunova, V. A. Kotovich, and V. A. Frank-Kamenetskii, Simultaneous crystallization of hexagonal cadmium selenide and ZnSe, InAs, and In_2Se_3, Zh. Tekh. Fiz., 25 : 2419 (1955).

35.   N. A. Goryunova, V. A. Kotovich, and V. A. Frank-Kamenetskii, X-ray diffraction investigation of the isomorphism of some gallium and zinc compounds, Doklady Akad. Nauk SSSR, 103 :659 (1955).

36.   N. A. Goryunova and V. D. Prochukhan, Solid solutions in quaternary systems based on InAs and InSb, Fiz. Tverd. Tela, 2 : 176 (1960).

37.   N. A. Goryunova, S. I. Radautsan, and V. I. Deryabina, Homogenization of alloys of the InAs−In_2Se_3 system by annealing under pressure, Fiz. Tverd. Tela, 1 : 512 (1959).

38.   N. A. Goryunova and S. I. Radautsan, Solid solutions in the InAs−In_2Se_3 system, Zh. Tekh. Fiz., 28 : 1917 (1958).

39.   N. A. Goryunova, S. I. Radautsan, and G. A. Kiosse, A new semiconducting compound in the In−Sb−Te system, Fiz. Tverd. Tela, 1 : 1858 (1959).

40.   V. S. Grigor'eva, Solid solutions in the Ga_2Te_3−In_2Te_3 system, Zh. Tekh. Fiz., 28 : 1670 (1958).

41. A. I. Gubanov and F. M. Gashimzade, Investigation of the symmetry of the electron energy bands in crystals of the $CdIn_2Se_4$ type, Fiz. Tverd. Tela, 1:1411 (1959).

42. A. I. Gubanov and F. M. Gashimzade, Structure of the energy bands in semiconductors of the $CdIn_2Se_4$ type, Fiz. Tverd. Tela, 2:255 (1960).

43. A. I. Gubanov and T. F. Mazets, Investigation of electrical conductivity of glassy semi-conductors of the As_2Te_3 type, Izv. Akad. Nauk SSSR, seriya fiz., 98:1276 (1964).

44. I. E. Gorshkov and N. A. Goryunova, Quasi-binary GaSb−InSb tie line of the gallium−indium−antimony system, Zh. Neorg. Khim., 3:668 (1958).

45. E. I. Elagina and N. Kh. Abrikosov, Investigation of $PbTe−Bi_2Te_3$ and $SnTe−Sb_2Te_3$ systems, Zh. Neorg. Khim., 4:1638 (1959).

46. V. V. Eremenko, Investigation of the spectral distribution of the photoconductivity of mixed $CdS_x−CdSe_{1-x}$ single crystals at 77 and 20°K, Fiz. Tverd. Tela, 2:2602 (1960).

47. B. A. Efimova and L. A. Kolomoets, Thermoelectric properties of PbTe−SnTe solid solutions, Fiz. Tverd. Tela, 7:424 (1965).

48. B. A. Efimova, T. S. Stavitskaya, L. S. Stil'bans, and L. M. Sysoeva, Mechanism of carrier scattering in some solid solutions based on lead and bismuth tellurides, Fiz. Tverd. Tela, 1:1325 (1959).

49. Yu. V. Zherdev and B. F. Ormont, Dependence of the forbidden bandwidth of ZnSe−CdSe phases on the structure and composition, Zh. Neorg. Khim., 5:239 (1960).

50. Yu. V. Zherdev and B. F. Ormont, Dependence of the forbidden bandwidth in the ZnSe−CdSe system on the structure and composition, Zh. Neorg. Khim., 5:1796 (1960).

51. I. P. Zvyagin, Problem of calculation of transport coefficients of heavily doped semi-conductors, Fiz. Tverd. Tela, 6:2972 (1964); Physics Letters, 11:5 (1964).

52. E. L. Zorina, Absorption of light in glassy As_2S_3 and As_2S_5 at the fundamental band edge, Fiz. Tverd. Tela, 7:331 (1964).

53. V. I. Ivanov-Omskii and B. T. Kolomiets, Some properties of InSb−GaSb alloys, Fiz. Tverd. Tela, 4:568 (1959).

54. B. I. Ivanov-Omskii and B. T. Kolomiets, Equilibrium solid solutions in the InSb−GaSb system, Fiz. Tverd. Tela, 1:913 (1959).

55. V. I. Ivanov-Omskii and B. T. Kolomiets, Dependence of the forbidden bandwidth on the composition of solid solutions in the InSb−GaSb system, Doklady Akad. Nauk SSSR, 127:135 (1959).

56. V. I. Ivanov-Omskii, B. T. Kolomiets, and A. A. Mal'kova, Optical and photoelectric properties of HgTe and its alloys with CdTe, Fiz. Tverd. Tela, 6:1457 (1964).

57. A. V. Ioffe and A. F. Ioffe, Influence of impurities on the thermal conductivity of semi-conductors, Doklady Akad. Nauk SSSR, 98:757 (1954).

58. A. V. Ioffe and A. F. Ioffe, Thermal conductivity of semiconductors, Izv. Akad. Nauk SSSR, seriya fiz., 20:65 (1956).

59. A. V. Ioffe and A. F. Ioffe, Thermal conductivity of semiconducting solid solutions, Fiz. Tverd. Tela, 2:781 (1960).

60. E. B. Ivkin, B. T. Kolomiets, and É. A. Lebedev, Measurement of the carrier mobility in glassy chalcogenide semiconductors, Izv. Akad. Nauk SSSR, seriya fiz., 28:1288 (1964).

61. G. V. Kokosh and S. S. Sinani, Influence of impurities on thermoelectric properties of $Sb_2Te_3−Bi_2Te_3$ solid solution, Fiz. Tverd. Tela, 1:89 (1959).

62. G. V. Kokosh and S. S. Sinani, Thermoelectric properties of alloys of the pseudo-binary system $Sb_2Te_3−Bi_2Te_3$, Fiz. Tverd. Tela, 2:1118 (1960).

63. B. T. Kolomiets and N. A. Goryunova, Electrical properties and structure of some materials in the Te−Sb−Se system, Zh. Tekh. Fiz., 25:984 (1955).

64. B. T. Kolomiets and Lin Chün-T'ing, Internal photoeffect spectrum of the ZnSe−CdSe system, Fiz. Tverd. Tela, 2:169 (1960).

65. B. T. Kolomiets and T. N. Mamontova, Internal photoeffect of molten chalcogenide glass, Doklady Akad. Nauk SSSR, 125:73 (1959).

66. B. T. Kolomiets and V. M. Lyubin, Properties and structure of ternary semiconducting systems. VI. Electrical and photoelectric properties of films of the $Sb_2S_3 - Bi_2S_3$ system, Fiz. Tverd. Tela, 1:740 (1959).

67. B. T. Kolomiets and A. A. Mal'kova, Absorption and photomagnetic effect spectra of $Cd_xHg_{1-x}Te$ solid solutions, Fiz. Tverd. Tela, 5:1219 (1963).

68. B. T. Kolomiets and T. F. Nazarova, Role of impurities in the conductivity of glassy As_2SeTe_2, Fiz. Tverd. Tela, 2:174 (1960).

69. B. T. Kolomiets, T. S. Stavitskaya, and L. S. Stil'bans, Investigation of thermoelectric properties of lead telluride and selenide, Zh. Tekh. Fiz., 27:73 (1957).

70. W. Kohn and J. M. Luttinger, Quantum theory of electrical transport phenomena, in: Problems in the Quantum Theory of Irreversible Processes (ed. V. L. Bonch-Bruevich) [Russian translation]. IL, Moscow (1961).

71. R. Kubo, Statistical mechanics of irreversible processes, in: Problems in the Quantum Theory of Irreversible Processes (ed. V. L. Bonch-Bruevich) [Russian translation]. IL, Moscow (1961).

72. V. N. Lange and A. R. Regel', Characteristic features of the electrical properties of continuous solid solutions in the Te−Se and Te−S systems, Fiz. Tverd. Tela, 1:562 (1959).

73. M. Loève, Probability Theory, 2nd ed. [Russian translation]. IL, Moscow (1962).

74. M. S. Mirgalovskaya and E. V. Skudnova, Investigation of alloys of the $AlSb - Al_2Te_3$ system, Zh. Neorg. Khim., 4:1113 (1959).

75. D. N. Nasledov and I. A. Fel'tin'sh, Problem of the electrical properties of gallium arsenoselenides, Fiz. Tverd. Tela, 1:565 (1959).

76. D. N. Nasledov and I. A. Fel'tin'sh, Electrical conductivity of gallium arsenoselenides at high temperatures, Fiz. Tverd. Tela, 2:823 (1960).

77. D. N. Nasledov, M. P. Pronina, and S. I. Radautsan, Some optical properties of solid solutions of indium arsenoselenides and arsenotellurides, Fiz. Tverd. Tela, 2:50 (1960).

78. E. N. Nikol'skaya and A. R. Regel', Formation of solid solutions and magnetic susceptibility of $HgTe - HgSe$, $HgTe - \beta - HgS$, $HgSe - \beta$-HgS systems, Zh. Tekh. Fiz., 25:1347 (1955).

79. E. N. Nikol'skaya and A. R. Regel', Some electrical properties of $HgTe - HgSe$, $HgTe - \beta$-HgS, $HgSe - \beta$-HgS solid solutions, Zh. Tekhn. Fiz., 25:1352 (1955).

80. I. M. Pilat, G. S. Borodinets, L. A. Kosyachenko, and V. I. Maiko, Some properties of the $CdSb - ZnSb$ system, Fiz. Tverd. Tela, 2:1522 (1960).

80a. A. I. Rezanov and S. I. Masharov, Theory of specific heat of substitutional solid solutions at low temperatures, Fiz. Metallov. i Metalloved., 13:3 (1962).

81. S. I. Radautsan, Investigation of the $InAs - In_2Se_3$ tie line of the $InAs - Se$ system, Zh. Neorg. Khim., 4:1121 (1959).

82. V. N. Romanenko and V. I. Ivanov-Omskii, Thermodynamics of solid solutions of some semiconducting systems, Doklady Akad. Nauk SSSR, 129:553 (1959).

83. S. S. Sinani and G. N. Gordyakova, $Bi_2Te_3 - Bi_2Se_3$ solid solutions as thermoelement materials, Zh. Tekh. Fiz., 26:2398 (1956).

84. A. G. Talybov, Electron-diffraction investigation of the structure of $SnSb_2Te_4$, Kristallografiya, 6:49 (1961).

85. V. G. Fomin and O. V. Bogorodskii, Investigation of microliquation in the crystallization of germanium−silicon alloys, Kristallografiya, 6:455 (1961).

86. G. I. Shmelev, Thermoelement materials based on ternary intermetallic compounds, Fiz. Tverd. Tela, 1:63 (1959).

87. A. D. Shneider and I. V. Gavrishchak, Structure and properties of the HgTe−CdTe system, Fiz. Tverd. Tela, 2 : 2079 (1960).

88. M. S. Abrahams, R. Braunstein, and F. D. Rosi, Thermal electrical and optical properties of (In, Ga)As alloys, J. Phys. Chem. Solids, 10 : 204 (1959).

89. A. Addamiano, The system ZnS−AlP, J. Electrochem. Soc., 107 : 1006 (1960).

90. A. Addamiano, On the preparation of the nitrides of aluminum and gallium, J. Electrochem. Soc., 108 : 1072 (1961).

91. G. R. Antell and D. Effer, Preparation of crystals of InAs, InP, GaAs, and GaP by a vapor phase reaction, J. Electrochem. Soc., 106 : 509 (1959).

92. T. Appel and S. W. Kurnick, Polaron band model and its application to Se−S semiconductors, J. Appl. Phys. 32(Suppl. to No. 10) : 2206-2210 (1961).

93. R. W. Armstrong, J. W. Faust, and W. A. Tiller, A structural study of the compound $AgSbTe_2$, J. Appl. Phys., 31 : 1954 (1960).

94. M. Avinor, Gold-activated (Zn, Cd)S phosphors, J. Electrochem. Soc., 107 : 608 (1960).

95. D. W. Ballentyne and B. Ray, Electroluminescence and crystal structure in the alloy system ZnS−CdS, Physica, 27 : 337 (1961).

96. R. Barrie and J. T. Edmond, Conduction band of InSb, J. Electronics, 1 : 167 (1955).

97. P. Baruch and M. Desse, Study of alloys of tin−indium antimonide, Compt. Rend., 241 : 1040 (1955).

98. L. C. Bennett and J. R. Wiese, Effects of doping additions on the thermoelectric properties of the intrinsic semiconductor $Bi_2Te_{2.1}Se_{0.9}$, J. Appl. Phys., 32 : 562 (1961).

99. H. Benel, Thermoelectric properties of antimony telluride and Sb_2Te_3−Bi_2Te_3 solid solutions, Compt. Rend., 247 : 584 (1958).

100. J. A. Beun, R. Nitsche, and M. Lichtensteiger, Photoconductivity in ternary sulfides, Physica, 26 : 647 (1960).

101. J. A. Beun, R. Nitsche, and M. Lichtensteiger, Optical and electrical properties of ternary chalcogenides, Physica, 27 : 448 (1960).

102. U. Birkholz, Investigation of the intermetallic compound bismuth telluride and its solid solutions with antimony and selenium regarding their applicability as materials for semiconductor thermal elements, Z. Naturforsch., 13a : 780 (1958).

103. J. A. Bland and S. J. Basinski, The crystal structure of Bi_2Te_2Se, Can. J. Phys., 39 : 1040 (1961).

104. R. Bowers, J. E. Bauerle, and A. J. Cornish, $InAs_{1-x}P_x$ as a thermoelectric material, J. Appl. Phys., 30 : 1050 (1959).

105. E. Braunersreuther, F. Kuhrt, and H. J. Lippmann, Hall constant and electron mobility of InSb, InAs, and $In(As_{0.8}P_{0.2})$ at high magnetic fields, Z. Naturforsch., 15a : 795 (1960).

106. R. Braunstein, A. R. Moore, and F. Herman, Intrinsic optical absorption in germanium−silicon alloys, Phys. Rev., 109 : 695 (1958).

107. H. Brooks, Theory of electrical properties of germanium and silicon, Advances Electronics and Electron Phys., 7 : 85 (1955).

108. G. Busch, P. Junod, E. Mooser, and H. Schade, The electric properties of $HgIn_2Te_4$, in: Semiconductors and Phosphors, Vieweg, Brunswick (1958), p. 470.

109. G. A. Busch and R. Kern, Die magnetischen Eigenschaften der $A^{III}B^V$-Verbindungen, Helv. Phys. Acta, 32 : 24 (1959).

110. G. Busch, E. Mooser, and W. B. Pearson, Neue halbleitende Verbindungen mit diamantähnlicher Struktur, Helv. Phys. Acta, 29 : 191 (1956).

111. G. Busch. H. J. Stocker, and O. Vogt, Magnetic susceptibility of silicon−germanium mixed crystals, Helv. Phys. Acta, 31 : 299 (1958).

112. G. Busch, H. J. Stocker, and O. Vogt, Magnetic susceptibility of charge carriers in germanium−silicon mixed crystals, Helv. Phys. Acta, 31 : 566 (1958).

113.   G. Busch and O. Vogt, Electrical conductivity and Hall effect of Ge—Si alloys, Helv. Phys. Acta, 33:437 (1960).

114.   G. Busch and O. Vogt, Magnetic susceptibility of Ge—Si mixed crystals, Helv. Phys. Acta, 33:889 (1960).

115.   G. Busch and U. Winkler, Elektrische Eigenschaften der intermetallischen Verbindungen Mg_2Si, Mg_2Ge, Mg_2Sn, und Mg_2Pb, Physica, 20:1067 (1954).

116.   J. Callaway, Model for lattice thermal conductivity at low temperatures, Phys. Rev., 113:1046 (1959).

117.   M. Cardona and H. S. Sommers, Effect of temperature and doping on reflectivity of germanium in the fundamental absorption region, Phys. Rev., 122:1382 (1961).

118.   R. P. Chasmar, E. W. Durham, and A. D. Stuckes, The thermal and electrical properties of cadmium and mercury tellurides, in: Proc. Internat. Conf. on Semiconductor Phys., Czech. Acad. Sci., Prague (1961), p. 1018.

118a.   F. Conforto, Abelsche Funktionen und algebraische Geometrie, Berlin (1966).

119.   G. J. Cosgrove, J. P. McHugh, and W. A. Tiller, Effect of freezing conditions on the thermoelectric properties of $BiSbTe_3$ crystals, J. Appl. Phys., 32:621 (1961).

120.   M. Culter, R. L. Fitzpatrick, and J. F. Leavy, The conduction band of cerium sulfide $Ce_{3-x}S_4$, J. Phys. Chem. Solids, 24:319 (1963).

121.   S. J. Czyzak, D. J. Craig, and C. E. McCain, Single synthetic cadmium sulfide crystals, J. Appl. Phys., 23:932 (1952).

122.   R. T. Delves, HgTe—MnTe alloys. II. Electrical properties, J. Phys. Chem. Solids, 24:885 (1963).

123.   R. T. Delves and B. Lewis, Zinc-blende type HgTe—MnTe solid solutions, J. Phys. Chem. Solids, 24:549 (1963).

124.   G. Destriau, Energy limit of the electroenhancement and electroextinction effects, Compt. Rend., 249:245 (1959).

125.   J. R. Drabble and C. H. L. Goodman, Chemical binding in bismuth telluride, J. Phys. Chem. Solids, 5:142 (1958).

126.   J. R. Drabble and H. J. Goldsmid, Thermal Conduction in Semiconductors, Pergamon Press, Oxford (1961).

127.   G. Dresselhaus, A. F. Kip, K. Hang-Ying, G. Wagoner, and S. M. Christian, Cyclotron resonance in germanium—silicon alloys, Phys. Rev., 100:1218 (1955).

128.   P. Duwez, R. H. Willens, and W. Klement, Continuous series of metastable solid solutions in silver—copper alloys, J. Appl. Phys., 31:1137 (1960).

129.   F. T. Dyson, The dynamics of a disordered linear chain, Phys. Rev., 92:1331 (1953).

130.   H. Ehrenreich, Electron mobility of indium arsenide phosphide $In-(As_yP_{1-y})$, J. Phys. Chem. Solids, 12:97 (1950).

131.   H. Fleischman, O. G. Folberth, and H. Preister, Halbleitende Mischkristalle vom Typ $A_{x/2}^{I}B_{1-x}^{IV}C_{x/2}^{V}$), Z. Naturforsch., 14a:999 (1959).

132.   P. D. Fochs, The measurement of the energy gap of semiconductors from their diffuse reflection spectra, Proc. Phys. Soc. (London), 69B:70 (1956).

133.   O. G. Folberth, Mischkristallbildung bei $A^{III}B^{V}$-Verbindungen, Z. Naturforsch., 10a:502 (1955).

134.   O. G. Folberth, Existence of tetrahedral phases, Z. Naturforsch., 14a:94 (1959).

135.   N. Fuschillo, J. N. Bierly, and F. J. Donahoe, Transport properties of the pseudo-binary alloy system $Bi_2Te_{3-y}Se_y$, J. Phys. Chem. Solids, 8:430 (1959).

136.   D. B. Gasson, P. J. Holmes, I. C. Jennings, J. E. Parrott, and A. W. Penn, The preparation and properties of In_2Te_3 and its alloys with InAs, in: Proc. Internat. Conf. Semiconductor Phys., Czech. Acad. Sci., Prague (1961), p. 1032.

137. J. H. Gisolf, The absorption spectrum of luminescent zinc sulfide and zinc—cadmium sulfide in connection with some optical, electrical, and chemical properties, Physica, 6:84 (1939).

138. M. Glicksman, Magnetoresistance of germanium—silicon alloys, Phys. Rev., 100:1146 (1955).

139. M. Glicksman, The galvanomagnetic effects in the germanium—silicon alloys, in: Semiconductors and Phosphors, Vieweg, Brunswick (1958), p. 452.

140. M. Glicksman, Mobility of electrons in germanium—silicon alloys, Phys. Rev., 111:125 (1958).

141. M. Glicksman, Galvanomagnetic effect in a semiconductor with two sets of spheroidal energy surfaces, Phys. Rev., 102:1496 (1956).

142. M. Glicksman and S. M. Christian, Conduction band structure of germanium—silicon alloys, Phys. Rev., 104:1278 (1956).

143. A. J. Goss, K. E. Benson, and W. C. Pfann, Dislocations at compositional fluctuations in germanium—silicon alloys, Acta Met., 4:332 (1956).

144. H. Guennoc, H. Anger, and M. Malard, Experimental study of semiconducting mixtures $Mg_2Sn_xPb_{1-x}$, in: Proc. Internat. Conf. on Semiconductor Phys., Czech. Acad. Sci., Prague (1961), p. 926.

145. G. Haacke and S. Poganski, Heat conductivity and electrical properties of the semi-conducting systems $AgSbTe_{2-x}Se_x$, in: Proc. Internat. Conf. on Semiconductor Phys., Czech. Acad. Sci., Prague (1961), p. 999.

146. H. Hahn, The effect of the choice of suitable components on the structures of ternary and quaternary phases. I. Structure of mixed crystals of InAs with InSe, In_2Se_3, and InTe, Naturwissenschaften, 44:534 (1957).

147. H. Hahn, G. Frank, W. Klingler, A. D. Störger, and G. Störger, Ternary chalcogenides. VI. Ternary chalcogenides of aluminium, gallium, and indium with zinc, cadmium, and mercury, Z. Anorg. Allgem. Chem., 279:241 (1955).

148. H. Hahn and D. Thiele, The effect of the choice of suitable components on the structures of ternary and quaternary phases. III. The systems In_2Se_3/InP, $In_2Se_3/InAs$, $In_2Te_3/InAs$, InSe/InAs, and InTe/InAs, Z. Anorg. Allgem. Chem., 303:147 (1960).

149. T. C. Harman and A. J. Strauss, Band structure of HgSe and HgSe—HgTe alloys, J. Appl. Phys., 32:2265 (1961).

150. I. J. Hegyi, S. Larach, and R. E. Shrader, Electroluminescence of Zn sulfoselenide phosphors with Cu activator and halide co-activators, J. Electrochem. Soc., 104:714 (1957).

151. S. T. Henderson, Band spectra of cathodo-luminescence, Proc. Roy. Soc. (London), A173:323 (1939).

152. S. T. Henderson, P. N. Ranby, and M. B. Halstead, Activation of ZnS and (Zn, Cd)S phosphors by gold and other elements, J. Electrochem. Soc., 106:27 (1959).

153. F. Herman, The electronic energy band structure of silicon and germanium, Proc. IRE, 43:1703 (1955).

154. F. Herman, Speculations on the energy band structure of Ge—Si alloys, Phys. Rev., 95:847 (1954).

155. F. Herman, M. Glicksman, and R. H. Parmenter, Semiconductor alloys, in: Progress in Semiconductors, London, Heywood (1957), Vol. II, Pt. I.

156. D. A. Jenny and R. Braunstein, Some properties of gallium arsenide—germanium mixtures, J. Appl. Phys., 29:596 (1958).

157. E. R. Johnson and S. M. Christian, Properties of Ge—Si alloys, Phys. Rev., 95:560 (1954).

158. R. E. Johnson and G. Towns, Graded energy gap alloy semiconductor $GaAs_xP_{1-x}$, J. Electrochem. Soc., 107:189 (1960).

159. E. Justi, Thermoelectric behavior of $Zn_xCd_{1-x}Sb$, in: Proc. Internat. Conf. on Semiconductor Phys., Czech. Acad. Sci., Prague (1961), p. 1074.

160. E. Justi, G. Neumann, and G. Schneider, Thermoelectric investigations of $Zn_xCd_{1-x}Sb$, Z. Physik, 156:217 (1959).

161. H. Kallman, B. Kramer, F. Spagnolo, and G. M. Spruch, Positive-hole motion and photovoltaic effects in zinc—cadmium sulfide phosphors, Phys. Rev., 123:1661 (1961).

162. M. Kikuchi and S. Iizima, Photoconductivity of Cd—ZnS mixed crystal, J. Phys. Soc. Japan, 15:357 (1960).

163. H. A. Klasens, On the nature of fluorescent centers and traps in zinc sulfide, J. Electrochem. Soc., 100:74 (1953).

164. H. Koelmans and C. G. Grimmeiss, The photoconductivity of $CdIn_2S_4$ activated with Cu or Au, Physica, 25:1287 (1959).

165. C. Kolm, S. A. Kulin, and B. L. Averback, Group III-V intermetallic compounds, Phys. Rev., 108:965 (1957).

166. W. Köter and B. Thoma, Aufbau ternärer Systeme von Metallen der dritten und fünften Gruppe des periodischen Systems, Z. Metallkunde, 46:293 (1955).

167. W. Köster and W. Ulrich, Zur Isomorphie der Verbindungen des Typs $A^{III}B^V$, Z. Metallkunde, 49:365 (1958).

168. K. Kreher, Die Bandstruktur von Galliumarsenid, Fortschr. Phys., 12:489 (1964).

169. A. A. Kremheller, A. K. Levine, and G. Gashurov, Hydrothermal preparation of two-component solid solutions from II-VI compounds, J. Electrochem. Soc., 107:12 (1960).

170. N. D. Lawson, S. Nielsen, E. N. Putley, and A. S. Young, Preparation and properties of HgTe and mixed crystals of HgTe—CdTe, J. Phys. Chem. Solids, 9:325 (1959).

171. A. Levitas, Electrical properties of germanium—silicon alloys, Phys. Rev., 99:1810 (1955).

172. A. Levitas, C. C. Wang, and B. H. Alexander, Energy gap of Ge—Si alloys, Phys. Rev., 95:846 (1954).

173. D. R. Mason and D. F. O'Kane, Preparation and properties of some peritectic semiconducting compounds, in: Proc. Internat. Conf. on Semiconductor Phys., Czech. Acad. Sci., Prague (1961), p. 1026.

174. J. P. McHugh, W. A. Tiller, S. E. Haszko, and J. H. Wernick, Phase diagram for the pseudo-binary system Ag_2Te—Sb_2Te_3, J. Appl. Phys., 32:1785 (1961).

175. J. W. Meclure, The electronic structure of rare earth monosulfides, J. Chem. Phys. Solids, 24:871 (1963).

176. M. Miksovsky, K. Smirous, and K. Toman, $Zn_xCd_{1-x}Sb$ solid solutions, in: Proc. Internat. Conf. on Semiconductor Phys., Czech. Acad. Sci., Prague (1961), p. 1087.

177. J. F. Miller, H. L. Goering, and R. C. Himes, Preparation and properties of AlSb—GaSb solid-solution alloys, J. Electrochem. Soc., 107:527 (1960).

178. S. Miyatani, Infrared absorption of α-$Ag_2Te_xS_{1-x}$, J. Phys. Soc. Japan, 14:1634 (1959).

179. E. Mooser and W. B. Pearson, The chemical bond in semiconductors. The group VB to VIIB elements and compounds formed between them, Can. J. Phys., 34:1369 (1956).

179a. T. Muto, Scient. Papers Inst. Phys. Chem. Soc. Res. (Tokyo), 34:377 (1938).

180. A. Nishiyama and T. Okada, Crystal structure of compounds $(SnSe)_{1-x-y}(SnTe)_x(PbTe)_y$, Mem. Fac. Sci. Kyushu Univ., B, 3:3 (1960).

181. L. Nordheim, Zur Electronentheorie der Metalle, Ann. Physik, 9:607 (1931).

182. A. Nussbaum, Electrical properties of pure tellurium and tellurium—selenium alloys, Phys. Rev., 94(2):337 (1954).

183. T. Okada, The electrical properties and crystal structure of IV_b-VI_b intermetallic compounds, J. Phys. Chem. Solids, 8:428 (1959).

184. R. H. Parmenter, Energy levels of a disordered alloy, Phys. Rev., 97:587 (1955).

185. R. H. Parmenter, Energy levels of a disordered alloy, Phys. Rev., 104:22 (1956).

186. R. H. Parmenter, Energy levels of a crystal modified by alloying or by pressure, Phys. Rev., 99 : 1759 (1955).

187. W. Paul, The effect of pressure on the properties of germanium and silicon, J. Phys. Chem. Solids, 8 : 196 (1959).

188. J. I. Petz, R. F. Kruh, and G. C. Amstutz, X-ray diffraction study of lead sulfide—arsenide glasses, J. Chem. Phys., 34 : 526 (1961).

189. J. C. Phillips, Optical absorption in germanium, J. Phys. Chem. Solids, 12 : 208 (1960).

190. F. Pingauli, Sensitization to x-rays of luminescent materials of the type ZnCdS : Mn by infrared radiation, Compt. Rend., 249 : 248 (1959).

191. E. S. Ritter and J. H. Schulman, Studies on the coprecipitation of cadmium and mercuric sulfides, J. Phys. Chem., 47 : 537 (1943).

192. H. Rodot, Semiconducteurs. Étude at propriétés du système $AgSb-Te_2-PbTe$, Compt. Rend., 249 : 1872 (1959).

193. M. Rodot, H. Rodot, and R. Triboulet, Some properties of $HgSe-HeTe$ solid solutions, J. Appl. Phys., 32 : 2254 (1961).

194. A. J. Rosenberg and A. J. Strauss, Solid solutions of In_2Te_3 in Sb_2Te_3 and Bi_2Te_3, J. Phys. Chem. Solids, 19 : 105 (1961).

195. F. D. Rosi, B. Abeles, and R. V. Jensen, Materials for thermoelectric refrigeration, J. Phys. Chem. Solids, 10 : 191 (1959).

196. J. Rupprecht and E. G. Maier, Neue Untersuchungen an halbleitenden Mischkristallen unter besonderer Berücksichtigung von Zustandsdiagrammen, Phys. Status Solidi, 8 : 3 (1965).

197. W. W. Scanlon, Recent advances in the optical and electronic properties of PbS, PbSe, PbTe, and their alloys, J. Phys. Chem. Solids, 8 : 423 (1959).

198. C. Shih and E. A. Peretti, The system $InAs-InSb$, J. Am. Chem. Soc., 74 : 608 (1953).

198a. K. L. Siegel, Analytic Functions of Several Complex Variables. Lectures delivered at the Institute of Advanced Studies (1948-1949).

199. K. Smirous and L. Stourac, Solid solutions of Bi_2Te_3 and Sb_2Te_3 as p-conducting materials for semiconductor thermoelements, Z. Naturforsch., 14a : 848 (1959).

200. K. Smirous and L. Stourac, Inversion of the type of conductivity in the semiconductor system $Zn_xCd_{1-x}Sb$, Z. Naturforsch., 14a : 1073 (1959).

201. I. Soudek, Effect irradiation on dielectric losses of luminescent zinc—cadmium sulfide, Czech. J. Phys., 7 : 119 (1957).

202. I. Soudek, Dependence of the spectral luminescence composition of zinc—cadmium sulfides on the excitation intensity, Czech. J. Phys., 8 : 336 (1958).

203. W. G. Spitzer and C. A. Mead, Conduction band minima of $Ga(As_{1-x}P_x)$, Phys. Rev., 133A : 872 (1964).

204. M. C. Stelle and F. D. Rosi, Thermal conductivity and thermoelectric power of germanium—silicon alloys, J. Appl. Phys., 29 : 1517 (1958).

205. H. Stöhr and W. Klemm, Über Zweistoffsysteme mit Germanium. I. Germanium—Aluminium, Germanium—Zinn und Germanium—Silicium, Z. Anorg. Allgem. Chem., 241 : 305 (1939).

206. L. Stourac, J. Tauc, and M. Lavetova, Electrical and optical properties of $Zn_xCd_{1-x}Sb$ solid solutions, in: Proc. Internat. Conf. on Semiconductor Phys., Czech. Acad. Sci., Prague (1961), p. 1091.

207. J. Tauc and A. Abraham, Optical investigation of the band structure of Ge—Si alloys, J. Phys. Chem. Solids, 20 : 190 (1961).

208. J. Tauc and A. Abraham, Reflection spectra of semiconductors with diamond and sphalerite structures, in: Proc. Internat. Conf. on Semiconductor Phys., Czech. Acad. Sci., Prague (1961), p. 375.

209. I. Teramoto and S. Takayanagi, Relations between the electronic properties and the chemical bonding of $Sb_x Bi_{2-x}Te_{3-y}Se_y$, J. Phys. Chem. Solids, 19 : 124 (1961).

210. C. Thurmond, Equilibrium thermochemistry of solid and liquid alloys of germanium and silicon. I. The solubility of germanium in elements of Groups III, IV, and V, J. Phys. Chem., 57 : 827 (1953).

211. F. A. Trumbore, Solid solubilities and electrical properties of tin in germanium single crystals, J. Electrochem. Soc., 103(11) : 97 (1956).

212. A. M. Tozen, Lattice thermal conductivity of germanium−silicon alloy single crystals at low temperatures, Phys. Rev., 122 : 450 (1961).

213. A. Washtel, ZnS : CuCl and (Zn, Cd)S : CuCl electroluminescent phosphors, J. Electrochem. Soc., 107 : 602 (1960).

214. A. Washtel, (Zn, Hg)S and (Zn, Cd, Hg)S electroluminescent phosphors, J. Electrochem. Soc., 107 : 682 (1960).

215. H. Weiss, Über die elektrischen Eigenschaften von Mischkristallen der Form $In(As_y P_{1-y})$, Z. Naturforsch., 11a : 430 (1956).

216. H. Weiss, Thermospannung und Wärmeleitung von III-V-Verbindungen und ihren Mischkristallen, Ann. Physik, 4 : 121 (1959).

217. G. Wendel, Zur Feldverstärkung bei ZnS−CdS−Mn-Phosphoren, Z. Naturforsch., 15a : 1010 (1960).

218. H. Winkler, H. Röppischer, and G. Wendel, Field intensification of ZnS−CdS : Mn phosphors under x-ray excitation, Z. Physik, 161 : 330 (1961).

219. A. Wold and U. Croft, Preparation and properties of the systems $LaTe_x Cr_{1-x}O_3$ and $LaFe_x Co_{1-x}O_3$, J. Phys. Chem., 63(3) : 447 (1959).

220. A. Wold and R. Arhott, Preparation and crystallographical properties of the systems $LaMn_{1-x}Mn_2O_{3+x}$ and $LaMn_{1-x}NiO_{3+x}$, J. Phys. Chem. Solids, 9(2) : 176 (1959).

221. J. C. Woolley and J. A. Evans, Temperature variation of optical energy gap for GaSb−InSb alloys, Proc. Phys. Soc. (London), 78 : 354 (1961).

222. J. C. Woolley, J. A. Evans, and C. M. Gillett, Optical energy-gap variation in the gallium antimonide−indium antimonide system, Proc. Phys. Soc. (London), 74 : 244 (1959).

223. J. C. Woolley and C. M. Gillett, Electrical properties of GaSb−InSb alloys, J. Phys. Chem. Solids, 17 : 34 (1960).

224. J. C. Woolley, C. M. Gillett, and J. A. Evans, Some electrical and optical properties of $InSb−In_2Te_3$ alloys, J. Phys. Chem. Solids, 16 : 138 (1960).

225. J. C. Woolley and P. N. Keating, Solid solubility of In_2Se_3 in some compounds of zinc-blende structure, J. Less Common Metals, 3 : 194 (1961).

226. J. C. Woolley and P. N. Keating, Some electrical and optical properties of $InAs−In_2Se_3$ and $InSb−In_2Se_3$ alloys, Proc. Phys. Soc. (London), 78 : 1009 (1961).

227. J. C. Woolley and D. C. Less, Equilibrium diagrams with InSb as one component, J. Less Common Metals, 1 : 192 (1959).

228. J. C. Woolley, C. M. Gillet, and J. A. Evans, Electrical and optical properties of GaAs−InAs alloys, Proc. Phys. Soc. (London), 77(3) : 700 (1961).

229. J. C. Woolley, Alloy semiconductors, in: Progress in Solid State Chemistry, Pergamon Press (1964), Vol. I.

230. J. C. Woolley, D. C. Less, and B. A. Smith, Equilibrium diagram of the $Ga_2Te_3−In_2Te_3$ system, J. Less Common Metals, 1 : 199 (1959).

231. J. C. Woolley and B. Ray, Solid solutions in $A^{II}B^{VI}$ tellurides, J. Phys. Chem. Solids, 13 : 151 (1960).

232. J. C. Woolley and B. Ray, Effects of solid solution of In_2Te_3 with $A^{II}B^{VI}$ tellurides, J. Phys. Chem. Solids, 15 : 27 (1960).

233. J. C. Woolley and B. Ray, Effects of solid solution of Ga_2Te_3 with $A^{II}B^{VI}$ tellurides, J. Phys. Chem. Solids, 16 : 102 (1960).

234. J. C. Woolley and B. A. Smith, Solid solutions in $A^{III}B^V$ compounds, Proc. Phys. Soc. (London), 72 : 214 (1958).

235. J. C. Woolley and B. A. Smith, Solid solution in zinc-blende type $A_2^{III}B_3^{VI}$ compounds, Proc. Phys. Soc. (London), 72 : 867 (1958).

236. J. C. Woolley, B. A. Smith, and D. G. Less, Solid solution in the GaSb−InSb system, Proc. Phys. Soc. (London), B69 : 1339 (1956).

237. D. A. Wright, Thermoelectric properties of bismuth telluride and its alloys, Nature, 181 : 834 (1958).

237a. P. Zallen and W. Paul, Band structure of gallium phosphide from optical experiment at high pressures, Phys. Rev., 134(6A) : 1628 (1964).

238. R. N. Zitter, InSb−Sn system, Trans. Met. Soc. AIME, 212 : 31 (1958).

SUPPLEMENTARY BIBLIOGRAPHY

239. N. A. Goryunova, V. S. Grigor'eva, P. V. Sharavskii, and L. A. Osnach, Solid solutions in the InAs−HgTe system, in: Proc. Twentieth Scientific Conference at the Leningrad Structural Engineering Institute, Leningrad (1962).

240. V. P. Chernyavskii, Semiconducting solid solutions of the system $mCuIn_2Te_2 \cdot (1 - m) \cdot 2CdTe$, in: Proc. Twentieth Scientific Conference at the Leningrad Structural Engineering Institute, Leningrad (1962).

241. A. V. Voitsekhovskii and N. A. Goryunova, Solid solutions in some quaternary semiconducting systems, in: Proc. Twentieth Scientific Conference at the Leningrad Structural Engineering Institute, Leningrad (1962).

242. A. S. Borshchevskii and D. N. Tret'yakov, Solid solutions between semiconducting compounds: indium and aluminum arsenides, in: Proc. Twentieth Scientific Conference at the Leningrad Structural Engineering Institute, Leningrad (1962).

243. M. I. Aliev and Kh. Ya. Khalilov, Investigations of the thermal conductivity, thermoelectric power, and electrical properties of the $InSb−ZnSnSb_2$ system, in: Proc. Second Scientific-Technical Conference at the Kishinev Polytechnical Institute, Physical Sciences Section (1966).

244. É. I. Gavrilitsa and I. K. Polushina, Investigation of the thermal conductivity of $(HgSe)_{3x}−(In_2Se_3)_{1-x}$ solid solutions, in: Proc. Second Scientific-Technical Conference at the Kishinev Polytechnical Institute, Physical Sciences Section (1966).

245. M. I. Alieva, M. A. Alieva, and A. A. Abdulrakhmanova, Investigation of the thermal conductivity of some $A^{III}B^V−A^{III}C^{VI}$ solid solutions, in: Proc. Second Scientific-Technical Conference at the Kishinev Polytechnical Institute, Physical Sciences Section (1966).

246. V. I. Bobrov, I. P. Molodyan, and S. I. Radautsan, Investigation of the phase diagram of the $(InSb)_x−(InTe)_{1-x}$ system, in: Proc. Second Scientific-Technical Conference at the Kishinev Polytechnical Institute, Physical Sciences Section (1966).

247. M. N. Ikizli and S. A. Tishkin, Investigation of the formation of solid solutions of gallium selenophosphides and tellurophosphides, in: Proc. Second Scientific-Technical Conference at the Kishinev Polytechnical Institute, Physical Sciences Section (1966).

248. L. I. Kleshchinskii, É. N. Khabarov, and P. V. Sharovskii, Determination of the limit of existence of solid solutions in the InAs−CdTe system, in: Proc. Twenty-Second Scientific Conference at the Leningrad Structural Engineering Institute, Leningrad (1964).

249. N. A. Goryunova and S. I. Radautsan, Solid solutions in the $InAs−In_2Te_3$ system, Doklady Akad. Nauk SSSR, 121:65 (1958).

250. S. I. Radautsan, Some semiconducting solid solutions based on indium arsenide, Ekspress-informatsiya pri Sovete Ministrov Mold.SSR(1960).

251. S. I. Radautsan, I. A. Madan, I. P. Molodyan, and R. A. Ivanov, Formation of solid solutions in the $InP-In_2Se_3$ system, Izv. Mold. Filiala AN, No. 3(69), p. 107 (1960).

252. S. I. Radautsan, Investigation of some complex semiconducting solid solutions and compounds based on indium, Czech. Phys. J., 12:382 (1962).

253. S. I. Radautsan, I. A. Madan, and R. A. Ivanova, Solid solutions of gallium selenophosphides, Izv. Akad. Nauk Mold.SSR, 10:98 (1961).

254. S. I. Radautsan and R. A. Ivanova, Formation of solid solutions based on complex compounds of the $A^{II}B^{IV}C^{IV}$ Type, Izv. Akad. Nauk Mold.SSR, No. 10 (1961).

255. S. I. Radautsan, V. V. Negreskul, and I. A. Madan, Some solid solutions based on a new compound In_4SbTe_3, Izv. Akad. Nauk Mold.SSR, 10:57 (1961).

256. I. P. Molodyan, S. I. Radautsan, and I. A. Madan, Some structural and thermal investigations of the compound In_4SbTe_3, Izv. Akad. Nauk Mold.SSR, 10:91 (1961).

257. S. I. Radautsan and I. A. Madan, Solid solutions of indium selenophosphides, Izv. Akad. Nauk Mold.SSR, 5:92 (1962).

258. N. A. Goryunova, S. I. Radautsan, and V. I. Deryabina, Method of homogenization of solid solutions of the indium arsenide—indium selenide system, Byulleten' izobretenii, No. 11 (1959).

259. I. P. Molodyan and S. I. Radautsan, Solid solutions based on indium antimonide in the indium—antimony—tellurium system, in: Investigations of Semiconductors, Izd. Kartya Moldovenyaské (1964), p. 143.

260. S. I. Radautsan and V. V. Negreskul, Solid solutions of gallium sulfophosphides, in: Investigations of Semiconductors, Izd. Kartya Moldovenyaské (1964), p. 158.

261. V. V. Negreskul and S. I. Radautsan, Some properties of solid solutions based on gallium phosphide, Izv. Akad. Nauk SSSR, seriya fiz., 28:1002 (1964).

262. Ya. L. Giller, V. E. Shmaevskii, and D. I. Vadets, Investigation of the pseudo-binary tie line $ZnSb-CdSb$ by the Debye method, Fiz. Metallov. i Metalloved., 11:311 (1961).

263. G. V. Nikitina and V. N. Romanenko, Solid solutions in the $In-Al-As$ system, Doklady Akad. Nauk SSSR, 170:107 (1966).

264. S. I. Radautsan and É. I. Gavrilitsa, Solid solutions in the $HgSe-In_2Se_3$ System, Izv. Akad. Nauk Mold.SSR, 10:95 (1961).

265. R. L. Stegman and E. A. Peretti, The ternary system indium telluride—indium antimonide—antimony, Chemical and Engineering Data, 11:496 (1966).

266. H. Krebs, K. Grün, D. Kallen, and W. Lippert, Mischkristallbildung zwischen den Metallen Arsen und Antimon und halbleitenden Chalkogeniden der vierten Hauptgruppe, Z. Anorg. Allgem. Chem., 308:200 (1961).

267. H. J. Naake and S. C. Belcher, Solid solutions in the system $Zn_3As_2-Cd_3As_2$, J. Appl. Phys., 35:3064 (1964).

268. C. C. Wang, M. Cardona, and A. G. Fischer, Preparation, optical properties, and band structure of boron monophosphide, RCA Rev., No. 2, p. 159 (1964).